Fiber Optics

Fiber Optics

ROBERT G. SEIPPEL

RESTON PUBLISHING COMPANY, INC.
A Prentice-Hall Company
RESTON, VIRGINIA

Library of Congress Cataloging in Publication Data

Seippel, Robert G.
Fiber optics.

Bibliography
Includes index.
1. Fiber optics. I. Title.
TA1800.S43 1983 621.36'92 83-13662
ISBN 0-8359-1970-6

1 3 5 7 9 10 8 6 4 2

Printed in the United States of America

Contents

Preface

High technology is not limited to lasers, computer integrated circuits, and other miniaturization. Fiber optics has a permanent position in this electronic revolution. It is no longer an infant in the electronics field. It has grown from a laboratory science into a practical tool for the transmission of information. Fiber-optic electronics is spreading rapidly to aerospace, heavy equipment, medicine, mining, oceanography, and most other technical fields.

Just what are fiber optics and why this rapid expansion? This new technology converts electrical impulses into light energy and transmits the light waves over an extremely small fiber to a receiver. At the receiver, the light energy is converted back to electrical impulses to perform its original task.

Why use fibers instead of electrical wire? First, because fibers are extremely light and small. Second, they are virtually immune to electrical interference. The fibers can be directed past high-voltage areas. They will neither radiate outward the signal they carry, nor are they susceptible to the acceptance of induced signals. Finally, optical fibers will accept a large bandwidth with extremely low power loss. Altogether, these attributes are indicative of efficient, advanced high technology.

Much of the credit for the phenomenal growth of fiber optics is due to the pioneers in the development of glass fibers, such as Standard Telecom Labs in England, Nippon Sheet Glass Company in Japan, and Bell Labs and Corning Glass Works in the United States. Since 1976, advancements in the fiber-optics field have come so quickly that it is difficult to determine who developed what. Dozens of companies in the United States are creating new fibers and connectors, new light launching devices, and related electronics. A momentous new technological breakthrough seems highly unlikely, simply because industry is already there. The future promises improved manufacturing techniques that will tend to bring prices down to meet the needs of consumer economics.

This book is written to provide the public with basic information on what fiber optics are, how they work, and how they are applied. To accomplish this task, the writer had to have the support of industry. The industries listed below are the major contributors to this effort. I would like to extend a special thanks for their help in compiling this fiber-optic story.

AT&T, New York, New York
Canoga Data Systems, Canoga Park, California
Corning Glass Works, Corning, New York
General Cable, Greenwich, Connecticut
General Telephone and Electronics Corp. Santa Monica, California
ITT Cannon Electric, Santa Ana, California
ITT Electro-Optical Products, Roanoke, Virginia
ITT Telecommunications, Raleigh, North Carolina
New York Telephone, New York, New York
Siecor Optical Cables Inc., Horseheads, New York
Western Electric, New York, New York

Robert G. Seippel, Ph.D.

I

Some Basics of Light Physics

An introduction to fiber optics must necessarily be preceded by some discussion on light. The discipline of fiber optics deals with light guided through tubes of glass or plastic by multiple reflections.

The history of light physics reaches back in time to the Greek thinkers. During this period there were two theories that were predominate, the tactile and the emission. The tactile theory suggested that the eyes sent out invisible antennae and were thus able to feel or sense things that were too distant to touch. The emission theory was that all objects emitted something and when entering the eye, the somethings were sensed.

Robert Hooke, a noted British scientist, published a book called *Micrographia* in 1665. In it he described light as small rapid vibrations and that variations in color are produced by these vibrations, thus originating the wave theory of light. Then in 1672, Sir Isaac Newton sent the society an account of his experiments in color separation with the prism. Shortly after the color theory, Newton developed his corpuscle theory.

As early as 1676, a Danish astronomer, Olaus Roemer, discovered that the speed of light was finite (measureable) and estimated its speed at 140,000 miles per second. It was not until 1926 at Mt. Wilson that Albert Michelson, an American, determined through more thorough experiments that the speed of light is 299,796 ± 4 kilometers per second.

Along with the fact that light has velocity, experiments by Christian Huygens, a seventeenth-century Dutch scientist, proved that light also had polarity (or direction) depending on the plane. However, it was not until the nineteenth century that detailed experiments were made with light reflection and refraction. The two findings (polarity and finite speed of light), along with several wave theories, further led to the theory of transverse waves. It was thought that an elastic medium pervaded all space and only changed when matter was in its way.

In other words, light was a thing that moved at a finite speed, was polarized, and could be modified by refraction or reflection. In his famous principle in 1690, Christian Huygens stated that each point of a wavefront may be regarded as a source of new waves. This was very much like Maxwell's later electromagnetic theory of light. James Clerk Maxwell, a Scot, through experiments with propagation of electromagnetic waves in the middle 1800s, derived mathematical constants for electricity and magnetism. By these constants he was able to establish a frequency and wavelength electromagnetic spectrum. The Maxwell mathematical constants of the electromagnetic wave theory allowed Heinrich Hertz, a German physicist, to produce wavelengths by electrical means in 1887.

In 1900, Max Planck found that light energy is emitted by atoms in multiples of an energy unit. The unit was called a quantum and its magnitude depended on wavelength. Seven years later, in 1907, Einstein found concentrations of energy, which he called photons, when light released electrons from atoms. They were propagated like particles from the electrons. In 1905, Einstein published the special theory of relativity. The importance of this publication is that the theory has some extremely striking practical implications. The first implication is that time and space separations are dependent on the selection of motion of the observer. The second prediction is that no body or physical effect travels faster than the speed of light. The third and probably the most known is the relation implication of mass m, energy E, and the speed of light c. The impact of the formula $E=mc^2$ is phenomenal. Technology has made great strides in areas such as atomic power and partical accelerators. Special relativity is by far the most practical and valid of all branches of physics.

Modern quantum mechanical light theory began around 1927 and incorporates the appropriate parts of the electromagnetic wave theory, the quantum theory, and the relativity theory. The progression of photoelectricity necessarily follows the progression of light. The probable beginning of photoelectricity was with the discovery of the photovoltaic cell by Alexandre Becqueral in 1839. Becqueral found that a battery he was working with developed a better charge when exposed to light.

In 1888, Heinrich Hertz working on the invisible portion of the spectrum revealed the fact that certain metals produced electrical effects when exposed to ultraviolet light. The next steppingstone to photoelectricity was Willoughby Smith's discovery that the resistance of selenium changes when light is applied. Among other scientists experimenting with this phenomena in the late nineteenth century was R. E. Day.

Of extreme importance in photoelectricity history and indeed science in general was the discovery of radiation in 1896 by Antoine Henri Becqueral, son of Alexandre Becqueral, the discoverer of the photovoltaic cell. The Becqueral family's work with light and electricity must be lauded as some of the greatest achievements of the time.

Just after the turn of the century, about 1905–1906, Max Planck developed his universal mathematical constant that light is emitted by atoms in multiples of a certain unit or quanta. The size of the unit being a quantum, which depends on the wavelength of the radiation. Einstein, in 1907, suggested that in order to give an adequate description of these localized pockets of energy developed from light exposure, that they had to be considered as particles separate from the atoms. Thus the concept of particles or photons of light energy was developed. In 1927, the photon theory was added to the quantum theory, the theory of relativity, and the electromagnetic theory, to establish the present-day light theory called quantum mechanics previously discussed.

Throughout the 1930s and 1940s many laboratories, such as Bell Telephone and General Electric, were developing photoelectric devices for commercial and military uses. The major breakthrough for the world of electronics was the invention of the transistor by Bardeen, Brattain, and Shockley of Bell Labs in 1948. This was the beginning of the age of electronics called *solid-state* or *semiconductor electronics*.

Since this discovery, electronics has been widely advanced by the design of integrated circuits. The principal designer was Kilby of Texas Instruments in 1958. Since integrated circuits are such a vast field in electronics today, we could not possibly give credit to everyone who is responsible for advancement in this area.

THE OPTICAL SPECTRUM

Optoelectronics is the branch of electronics that deals with light. Electronic devices involved with light operate within the optical part of the electronic-frequency spectrum. The visible range is only a narrow portion of the optical spectrum. There are three basic bands of the optical frequency spectrum. These are:

1. *Ultraviolet:* band of light wavelengths that is too short for the human eye to see.

2. *Visible:* band of light wavelengths that the human eye responds to.
3. *Infrared:* band of light wavelengths that is too long to be seen by the human eye.

Before we discuss each of the three basic bands of the optical spectrum, we shall explain several things that are involved in it.

A wavelength λ (the lowercase Greek letter lambda) is the amount of space occupied by the progression of an electromagnetic wave. Calculation of the size of the wavelength is dependent on the frequency of the wave f and the velocity of light c.

$$\lambda = \frac{c \text{ (velocity of light)}}{f \text{ (frequency)}}$$

$$\lambda = \frac{c}{f}$$

The frequency of the wave f is the same frequency that originated the radiated wave. The velocity of the wave c is the velocity of light, which is approximately the same velocity for all electromagnetic waves, that is, 300,000,000 meters per second (186,000 miles per second). The velocity of light varies slightly in different materials.

An audio frequency of 15,000 hertz (Hz; cycles per second) would have a wavelength of

$$\lambda = \frac{300{,}000{,}000}{15{,}000} = 20{,}000 \text{ meters (m)}$$

A radio frequency of 300 megahertz (MHz) would have a wavelength of

$$\lambda = \frac{300{,}000{,}000}{300{,}000{,}000} = 1 \text{ m}$$

Larger frequencies such as 3000 MHz would have a wavelength of 0.1 m.

Even larger frequencies such as 300,000 MHz would radiate a wavelength of 0.001 m, which could also be written as 1000 micrometers. It is convenient to discuss smaller wavelengths in micrometers because of the large size of the frequency involved.

1 micrometer (μm) = 10,000 angstroms (Å)

In terms of scientific notation,

$$1 \text{ angstrom } (\text{Å}) = 10^{-10} \text{ m}$$

$$1 \text{ micrometer } (\mu\text{m}) = 10^{-6} \text{ m}$$

The optical spectrum operates at wavelengths of 0.005 to 4000 μm (50 to 40,000,000 Å). In frequencies, these are extremely high values (6×10^{16} to 7.5×10^{10} Hz).

Ultraviolet Wavelengths

Wavelengths for ultraviolet radiated frequencies range from 0.005 to 0.3900 μm (50 to 3900 Å). In frequencies, these are 6×10^{16} to 7.69×10^{14} Hz.

Visible Wavelengths

Visible radiated wavelengths range from 0.3900 to 0.7500 μm (3900 to 7500 Å). In frequencies, these are 7.69×10^{14} to 4×10^{14} Hz.

Infrared Wavelengths

Wavelengths for infrared radiated frequencies range from 0.7500 to 4000 μm (7500 to 40,000,000 Å). In frequencies, these are 4×10^{14} to 7.5×10^{10} Hz. Infrared frequencies are too low and ultraviolet frequencies are too high for human eye response.

Infrared wavelengths are too long for human-eye response and ultraviolet wavelengths are too short for human-eye response. Another point worth mentioning is that the wavelength range for infrared (0.75 to 4000 μm) is much greater than either visual or ultraviolet wavelength ranges.

Color

The human eye sees violet (approximately 0.43-μm wavelength) on one side of the color spectrum and red (approximately 0.68-μm wavelength) on the opposite side of the spectrum. In between these extremes, the eye sees blue, green, yellow, and orange. Two beams of light that have the same wavelength are seen as the same color. Two beams that are seen to have the same color usually have the same wavelength. However, a mixture of two colors such as green and red will match a beam of blue. Therefore, the wavelengths are not necessarily the colors they may seem to the eye. Matches of colors are only correct to the person who has extremely good vision.

The graduation of colors and their wavelengths from one end of the color (visual) spectrum have overlapping wavelengths depending on the observer. Table 1–1 shows typical colors along with their estimated wavelengths and radiated frequencies.

TABLE 1–1 Optical Colors and Their Wavelengths

	Color	*Wavelength (μm)**	*Frequency (Hz)*
	Ultraviolet	0.005–0.39	6×10^{16}–7.69×10^{14}
VISUAL RESPONSE	Violet	0.40 –0.45	7.5 –6.6 $\times 10^{14}$
	Blue	0.45 –0.50	6.6 –6.0 $\times 10^{14}$
	Green	0.50 –0.57	6.0 –5.27 $\times 10^{14}$
	Yellow	0.57 –0.59	5.27–5.01 $\times 10^{14}$
	Orange	0.59 –0.61	5.01–4.92 $\times 10^{14}$
	Red	0.61 –0.70	4.92–4.28 $\times 10^{14}$
	Infrared	0.70 –4000	4.28×10^{14}–7.5×10^{10}

*Approximates only, with overlapping wavelengths and frequencies.

SOME "LIGHT" PHYSICS

Energy States

Each atom has several energy states (see figure 1–1). The lowest energy state is the ground state. Any energy state above ground represents an excited state and can be called by names such as first or second energy states or intermediate energy states. If an atom is in one of its energy states (E_4) and decays to a lower energy state (E_2), the loss of energy (in electron volts) is emitted as a photon. The energy of the photon is equal to the difference between the energy of the two energy states. The process of decay from one energy state to another is called *spontaneous decay* or *spontaneous emission*.

An atom may be irradiated by some light source whose energy is equal to the difference between ground state and some energy state. This

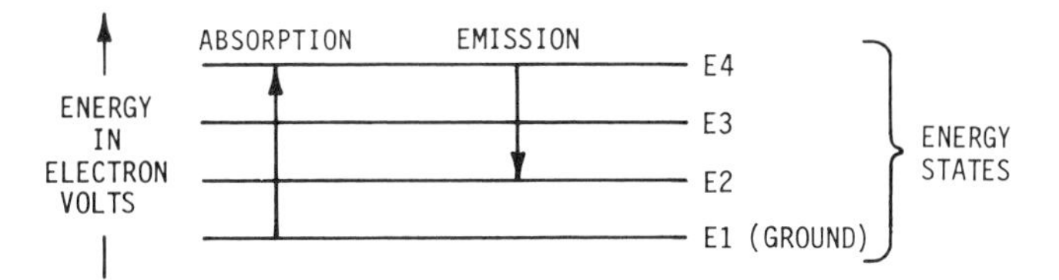

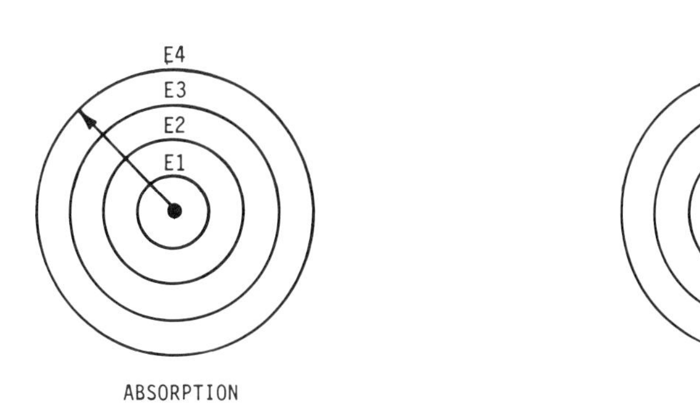

Figure 1–1 Energy States in Atoms

can cause an electron to change from one energy state to another by absorbing light energy. This transition from one energy level to another is called *absorption*. In making the transition the atom absorbs a packet of energy called the photon. This is similar to the process of emission.

The energy absorbed or emitted (photon) is equal to the difference between the energy states. For instance, $E_2 - E_1 = E_p$ (energy of photon).

$$E_2 - E_1 = E_p \text{ (energy of photon)}$$

Also $$E_p = hf$$

where h = Planck's constant

f = frequency of the radiation

The value of h has a dimension of energy and time. The value is 6.625×10^{-34} joule-second. The value of f is in hertz. Photon energy is said to be directly proportional to frequency.

Photon energy may also be expressed in terms of wavelength. You may recall that wavelength is equal to the speed of light in meters per second divided by frequency,

$$\lambda = \frac{c}{f}$$

where λ = wavelength

c = speed of light, meters per second

f = frequency of the electromagnetic wave, hertz

The formula can also be stated in the following manner:

$$c = f\lambda \qquad \text{and} \qquad f = \frac{c}{\lambda}$$

If we substitute this value of f in the formula for the energy of a photon, we have a second formula that relates the energy of the photon to wavelength.

$$E_p = hf$$

$$E_p = h\left(\frac{c}{\lambda}\right)$$

$$E_p = \frac{hc}{\lambda}$$

Propagation Sources

Light rays on beams are propagated at various wavelengths. Some sources radiate a much narrower bandwidth than others. Monochromatic light is radiation that has one wavelength (or an extremely narrow band of several wavelengths). Heterochromatic light is radiation that has several very distinct wavelengths. Panchromatic light is radiation that includes wavelengths in a great range of the spectrum.

Action of Light on Material

All material does not ''see'' light in the same manner. There are basically three different light-receptor materials. These are opaque, transparent, and translucent. Material that does not allow any light to travel through it is called *opaque*. If you can see light partially through a substance but not clearly, the substance is called *translucent*. Finally, *transparent* material allows you to see through it clearly.

Photons

Response of the human eye to light is determined by the energy of the radiation. Radiant energy travels in energy groups known as *photons*. The frequency of the radiation determines the strength of the photons. The higher the frequency of the radiation, the higher the energy of the photons. As an example, wavelengths in the visible range of the spectrum near the ultraviolent range have short wavelengths, higher frequencies, and higher photon energy. Furthermore, wavelengths in the visible range of the spectrum near the infrared range have long wavelengths, lower frequencies, and lower photon energy. When an electron is freed by absorbing the energy of a photon the result of the combination is called a *photoelectron*. Again the energy of the photoelectron is dependent on the frequency of the radiation. The energy of a photon is expressed in watts and can be calculated by the following equation:

$$W = hf$$

where W = energy, watts

h = Planck's constant (6.63×10^{-34} joule-second)

f = frequency of the photon wavelength, hertz

Diffraction (See Figure 1–2)

Up to this point we have assumed that light travels in straight lines. This is not altogether true. In experiments by Newton called *Newton's rings,* he discovered that light does spread, though to a very small extent, from the

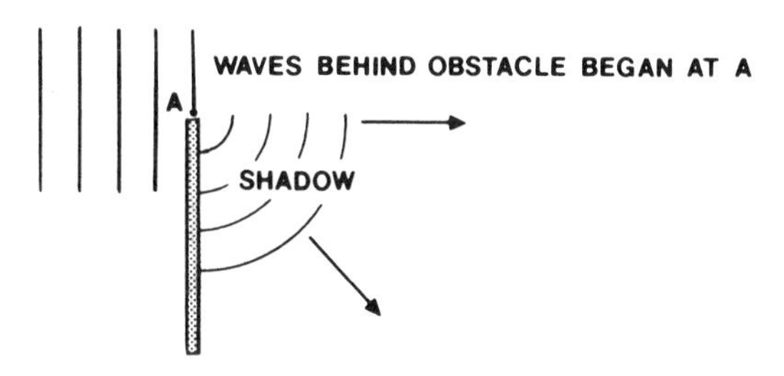

Figure 1–2 Diffraction

edges of the light beam. The shadow of an opaque obstacle is not perfectly sharp, but some light does indeed penetrate into the shadow area. A portion of the wave is stopped, however, since each point of the wavefront is in fact a source of waves. These waves bend around the obstacle. This property of bending around obstacles is called *diffraction*. We might add that light does not always correspond to its visual appearance.

Light Reception

The reception of light is completely dependent on what use or application is intended. If the desire is to light a room, the light should reflect diffusely from most objects in the room with absorption dependent on the materials and colors in the room. There is little or no use for refractive light. If the light were to be used for fiber-optic applications, fiber ends would reflect at zero percent if possible and the light refracted into the fiber would be maximum. The amount of absorption within the fiber would be held to a minimum. If the light were to be used for photodetection the amount of absorption would be extremely high (100 percent if possible). The amount of reflection would be held to a minimum. Thus the application decides the use of materials that are low or high absorbers, reflect or do not reflect, and refract or do not refract.

Surface Reflection (See figure 1–3.) Whenever a beam of light from one medium, such as air, strikes a second mirrorlike medium such as glass, part of the beam is reflected and the other part is refracted. If the surface is uneven, the reflection is called *diffuse reflection*. If the surface is smooth, such is the case with optical fibers, the reflection is called *specular reflection*. The angle that the incident rays strike the second medium is called the *angle of incidence*. This angle is the angle made by the incident beam and a line normal (perpendicular) to the boundary of the two media. Part of the incident beam is reflected at an angle called the *angle of reflection*. This angle is made by the reflected beam and a line normal (perpendicular) to the boundary of the two media:

Angle of incidence = Angle of reflection

If the angle of incidence varies, so does the angle of reflection by the same amount. The angle of incidence, the angle of reflection, and the angle of refraction all lie in the same plane.

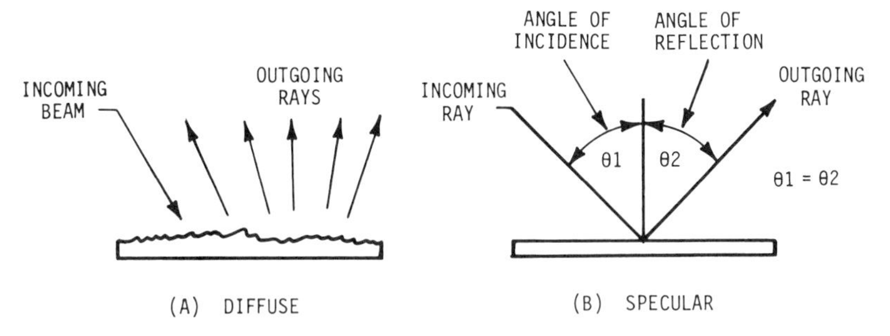

Figure 1–3 Reflection Types

Fresnel Reflection Power losses are those that are due to the difference in the index of refraction between the air and the fiber at the fiber end. This difference results in reflections called *Fresnel losses*. Variation of Fresnel reflections are related to unpolarized light, light polarized perpendicular to the plane of incidence, and light polarized parallel to the plane of incidence.

Surface Refraction (See figure 1–4.) Whenever a beam of light passes from one medium, such as air, to another medium, such as water, the

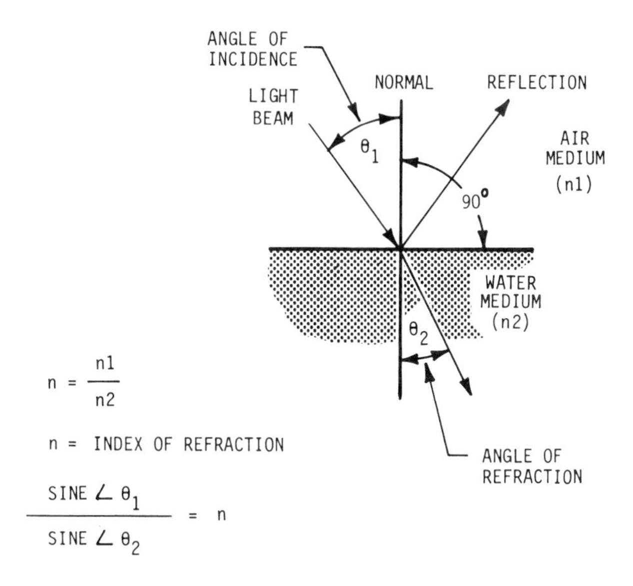

Figure 1–4 Refractive Indexes and Angles

beam separates at the intersection. Part of the beam is reflected back into the incident medium (air) and part is refracted into the second medium (water). The angle made by the incident beam and a line normal (perpendicular) to the intersection is called the *angle of incidence*. The angle made by the reflected beam and the line normal to the intersection is called the *angle of reflection*. The angle of incidence and the angle of reflection are equal. The angle made by the refracted beam and a line normal (perpendicular) to the intersection is called the *angle of refraction*.

The ratio of the indexes of refraction of the two indexes of refraction is thus

$$n = \frac{n_2}{n_1}$$

or

$$n = \text{index of refraction}$$

The ratio of the sine of the angle of incidence to the sine of the angle of refraction is equal to the index of refraction:

$$n = \frac{\sin \angle \theta_1}{\sin \angle \theta_2}$$

Therefore,

$$\frac{n_2}{n_1} = \frac{\sin \angle \theta_1}{\sin \angle \theta_2}$$

or

$$n_1 \sin \angle \theta_1 = n_2 \sin \angle \theta_2$$

Furthermore, the ratio is a constant ratio. Whenever the angle of incidence changes, the angle of refraction changes to retain the ratio. The law is *Snell's law*. Snell was a Dutch astronomer and professor of mathematics at the University of Leyden in Holland.

The index of refraction of a medium is the ratio of the speed of light and its velocity in a particular medium. Refer to table 1–2 for typical indexes of refraction *(n)*.

TABLE 1–2 **Indexes of Refraction**

Substance	n	*Substance*	n
Air	1.0003	Glass, flint	1.63
Benzene	1.50	Glycerin	1.47
Carbon disulfide	1.63	Ice	1.31
Diamond	2.42	Quartz	1.46
Ethyl alcohol	1.36	Water	1.34
Glass, crown	1.52		

Total Internal Reflection and Glass Piping of Light (See figure 1–5.) For a great number of years it has been known that total internal reflection can occur when light passes through a medium such as air into a medium such as glass. This may be illustrated as a glass pipe, which makes possible the transfer of light from one point to another through a series of internal reflections. The reader can see the progression. The amount of light that emerges is almost the same as that entering.

Absorption Whenever a beam of light enters matter its intensity decreases as it travels farther into the medium. There are two types of ab-

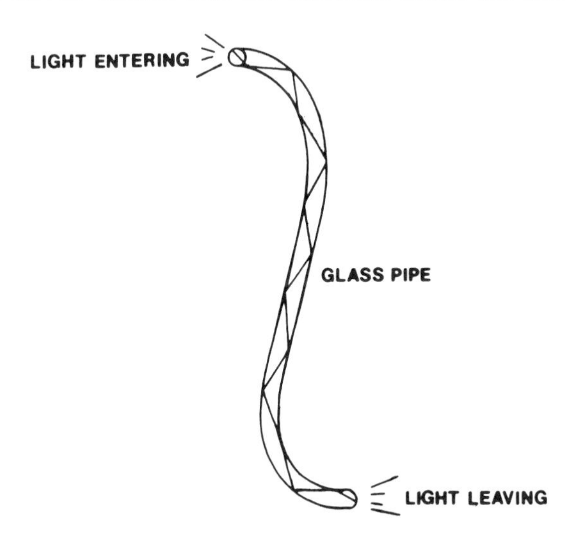

Figure 1–5 Glass Piping of Light

sorption, general and selective. *General absorption* is said to reduce all wavelengths of the light by the same amount. There are no substances which absorb all wavelengths equally. Some material such as lampblack absorbs nearly 100% of light wavelengths. With *selective absorption* the material selectively absorbs certain wavelengths and rejects others. Almost all colored things such as flowers and leaves owe their color to selective absorption. Light rays penetrate the surface of the substance. Selective wavelengths (therefore colors) are absorbed, whereas others are reflected or scattered and escape from the surface. These wavelengths appear in color to the human eye. In optoelectronics it is imperative that nearly 100% of incident beams be absorbed in devices such as solar cells and detectors. On the reverse side, 0% absorption is important in the transmission of light waves in an optical fiber cable and lenses.

Scattering Scattering is differentiated from absorption in the following manner. With true absorption, the intensity of the beam is decreased in calculable terms as it penetrates the medium. Light energy absorbed in the material is converted to heat motion of molecules. Consider a long tunnel where you could only see light from one end. As you walked nearer that end the light gets brighter. The light is absorbed as it travels through the tunnel at the absorption rate of the air medium. If the tunnel were then filled with a light cloud of smoke, the smoke would scatter some of the light from the main beams; therefore, the intensity of the light from a fixed distance would decrease. You may observe scattering effects by watching dust particles as the sun shines in a window. Parts of the rays are scattered by the dust particles.

A scattering type known as *Rayleigh scattering* is caused by micro irregularities in the medium. A wave passing through the medium strikes these micro irregularities in the mainstream of the waves. The waves reflected from the microparticles are spherical and do not follow the main wave, but scatter. Therefore, the intensity of the beam is diminished. In fiber-optic application the composition of the glass must be considered to ensure low Rayleigh scattering. Silica has low scattering losses.

Polarized Light Waves (See figure 1–6.) Light waves are either polarized or unpolarized. If the waves oscillate randomly through space or material and have no general direction, they are said to be *unpolarized*. If the waves oscillate at the same amplitude and trace out a circle in a plane perpendicular to the direction of propagation, they are said to be *polarized*. If the waves oscillate at the same amplitude along a line in a plane perpendicular to the direction of propagation, the waves are said to be linearly or *plane polarized.* There are other polarizations, such as the elliptically polarized wave.

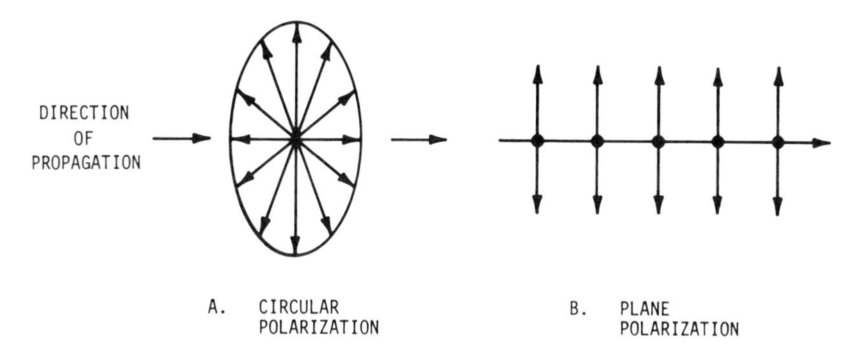

Figure 1–6 Polarization of Waves

A beam of unpolarized light may be the result of two beams combined in two different planes at two different amplitudes with no phase relationship. The directional variations of the two waves are not related nor are their magnitudes.

Coherency (See figure 1–7.) In order to have coherency the waves must have a consistent relationship between their wave troughs and wave

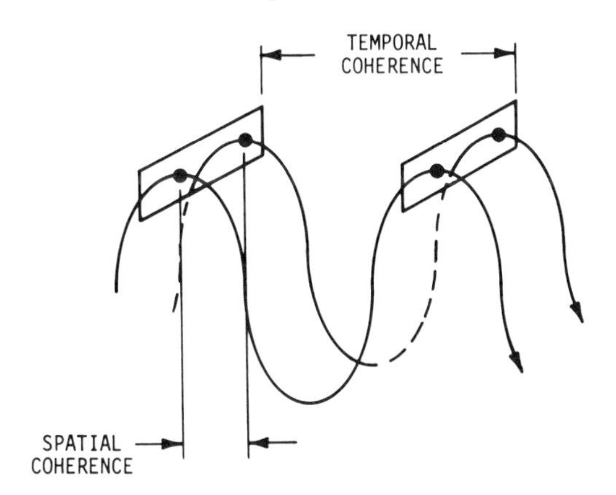

Figure 1–7 Coherence

crests. Wave *troughs* are points of minimum vibration and wave *crests* are points of maximum vibration. Coherency is divided into two major relationships, spatial and temporal. To be *spatially coherent,* waves must have phase correlation across a wavefront at any point in time. Spatially coherent waves also maintain their shape in relation to time. To be *temporally coherent,* waves must be frequency constant so that phase correlation is in the direction of propagation.

Dispersion Whenever a beam of light enters matter its velocity decreases within that medium. The subject of dispersion deals with the speed of light in a medium and the variation in conjunction with wavelength. The speed of light in a medium can be calculated by its refractive index. The *refractive index (n)* of any material is the ratio of the speed of light in a vacuum to its speed in the medium:

$$n = \frac{c}{v}$$

where n = refractive index of the medium
c = speed of light in a vacuum
v = speed of light in a medium

Then the speed of light (v) in the medium may be calculated by

$$v = \frac{c}{n}$$

The refractive index for air is 1.000, whereas optical glasses have values of 1.520 to 1.720 for refractive indexes. Any change then in refractive index would represent a change in velocity.

The dispersion of each material is different. Curves are supplied by manufacturers that plot wavelength against refractive index. As the wavelength decreases the refractive index increases and the dispersion $(dn/d\lambda)$ increases. The rate of increase becomes greater at shorter wavelengths. *Dispersion* is the spreading of light rays. In the propagation of light rays through a medium such as an optical fiber, the problems of dispersion are multiplied. With fiber optics there are two types of dispersion, intermodal and intramodal. *Intermodal* dispersion is caused by the propagation of rays of the same wavelength along different paths through the fiber medium. This results in the wavelengths arriving at the opposite end of the fiber at different times. *Intramodal* dispersion is due to variations of the refractive index of the material that the fiber was made from. Dispersion is measured in nanoseconds per kilometer (ns/km) in fiber-optic applications.

2

Introduction to Fiber Optics

Fiber optics is the science that deals with the transmission of light through extremely thin fibers of glass, plastic, or other transparent material. Optical fibers are dielectric waveguides for electromagnetic energy at optical wavelengths. The fibers provide a path for a single beam of light or in multiples, such as the transposition of a complete image. The fibers are provided as a single fiber or a cable bundle. They may be bent or curved (within limits) to meet the needs of special routing.

HISTORY OF FIBER-OPTIC ELECTRONICS

Probably the founder of fiber-optic electronics was the British physicist John Tyndall. Tyndall found in 1870 that light will follow water into a container and flowing from the container. It was not, however, until after World War II and into the 1950s that major corporations began study in the area of light and fiber optics. In 1968, Standard Telecom Labs in England developed silica glass with an extremely low transmission loss. Since then the Corning Glass Works and Bell Telephone Labs in the United States and Nippon Sheet Glass Company in Japan have developed glass fibers with typical loss factors of 1 to 2 decibels per kilometer (dB/km) and even much lower (0.1 dB/km) under lab conditions. These low-loss capabilities lend themselves to the development of low-cost, high-efficiency fiber-optic communication systems. After 1976, advancements have come so quickly that it is difficult to determine who developed what. Dozens of companies in the United States are creating new fibers and connectors, new light launching devices, and related electronics. A momentous new technological breakthrough seems highly unlikely, simply because industry is already there. The future looks to manufacturing techniques that will bring prices down to meet the needs of consumers.

OPTICAL FIBER FREQUENCY

Most optical fibers are manufactured for use in the visible frequency spectrum. Frequency responses tend to fall off on the ultraviolet and the infrared ends of the spectrum. Modern fiber optics have been manufactured to operate in the 4000-Å-wavelength range on the ultraviolet range of spectrum. On the infrared end of the spectrum, wavelengths of 10 μm are now common.

WHY USE OPTICAL FIBERS FOR ELECTRONIC COMMUNICATION?

The advantages of using fiber optics in communication networks are almost as limitless as a designer's ability to create. The optical transmission path provides electrical input and output isolation. To a designer this means total freedom from ground loops, along with lightning-safe installation. Bandwidths are not a problem because they are independent of the cable size. Extremely long distance cables are available, along with an endless variety of bandwidth products. This means greater data rates at longer distances than wire or coaxial cabling.

One of the major problems encountered in communication electronics is electromagnetic induction (EMI). With fiber optics there is no EMI susceptibility, no induced noise, and no crosstalk. Finally, the fibers and cables are extremely lightweight and low in cost compared to wire. These reasons add up to lower cost in installation and less maintenance time.

FIBER OPTICS IN OPERATION

In cooperative experiments with Bell Labs in 1973, Western Electric, in its Atlanta, Georgia, facility designed and developed light-guide fibers for telecommunication cabling. The fibers were arrayed into ribbons and placed in experimental cables at Atlantic Bell Labs. These cables were shipped to Chicago for testing. In May 1978, AT&T reported a successful one-year trial of a full-service light-wave installation in Chicago. The Chicago test involved a 1.5-mile link under downtown streets between two Illinois Bell Telephone switching stations and between one of the stations and an office housing a number of customers.

In late 1979, AT&T set three more installations into operation. A Connecticut hookup used light-wave technology in a portion of telephone network between the telephone company central office and local terminals located near customer premises. At about the same time a Florida installation provided service to a power station. In Arizona, Mountain Bell supplied service to a government agency near a power station. These experimental stations provided proof that light-wave systems are immune to electrical interference.

Probably New York Telephone set up the most comprehensive and innovative telecommunication systems to date. Two major light-wave installations began operating in 1980. The temporary one in upstate New York was a 2½-mile light-wave link that operated successfully at Lake Placid during the Winter Olympic Games. The second, a permanent installation, links telephone company offices in White Plains with those on East 38th Street in downtown Manhattan. These systems carry incredibly large volumes of information, such as voice, video, and computer data and in substantially less space than standard cable and wire systems. Both projects involved Bell Telephone Laboratories, AT&T, Western Electric, and New York Telephone.

In October 1978, General Cable Corporation announced that it would furnish the first optical fiber cable for use with the railroad industry at Union Pacific's Denver, Colorado, terminal. The optical fiber cable was designed and manufactured by General Cable to link a remote CCTV video camera with terminal office equipment to record information, car numbers, and reporting marks printed on passing freight cars. This information is used to check the accuracy of computerized advance makeups of arriving and departing trains, thereby verifying the makeup of the trains as well as the location of all the cars in the trains. The television also provides a check of all cars received from and sent to other railroads in the area. By having accurate information available promptly, switching can be prearranged and speeded up, thereby improving service to shippers. The optical fiber cable is an aerial installation and was continuously monitored for a year to assess its performance under conditions of vibration, ice and wind loading, and significant temperature changes. The optical fibers were manufactured by Corning Glass Works, Corning, New York.

Not only must the transmission medium be able to generate reliability over a broad range of harsh environment with a minimum of maintenance but it must also be economically feasible and, ideally, be immune to electrical and electromagnetic interference as well as offer a broad bandwidth of transmission capacity available for other railroad services.

Another optical fiber cable was designed, manufactured, and installed by General Cable for the U.S. Air Force's Arnold Engineering Development Center in Tennessee. The cable connects rocket engine test sites with a central data processing facility for real-time data analysis at the free world's largest rocket test facility. This was the first optical fiber cable fully engineered with splices, terminals, and other components for a working high-speed data transmission link—up to 150 megabits per second (Mbps).

For NASA, General Cable designed, manufactured, and installed at Cape Kennedy an optical fiber cable for transmission of high-speed digital data and video signals between locations at the center. They also designed, developed, and manufactured for General Telephone & Electronics Corporation, the world's first optical fiber cable to provide regular telephone service to the public. It was placed in operation in California. Also, a similar optical fiber cable providing telephone service in Brussels, Belgium, was manufactured by General Cable.

Teleprompter Cable Television Incorporated hooked up an 800-ft fiber optic link from an antenna on a Manhattan roof to head-end equipment on the bottom floor of a 34-floor building. The U.S. Navy's Airborne Light Optical Fiber Technology (ALOFT) program successfully demonstrated the use of fiber optics in flight tests on board an A-7 test aircraft. Point-to-point optical links were used. Northrup Corporation and ITT Cannon Electric on a joint research program developed a fiber-optic bus to interconnect avionics equipment on board an F-5E tactical fighter. The bus is used on the F-16 and F-18 aircraft.

For several years now a fiber-optic telephone system has operated with little or no failure on the USS *Littlerock*. Sonar links are now installed on U.S. Navy submarines. The U.S. Army is using optical fiber cables in ground tactical and strategic telecommunications (deployable). General Motors has developed a fiber-optic harness system to transmit a control signal to dashboard instruments. Sperry Univac and Digital Equipment Corporation are using fiber optics to connect mainframes and peripheral equipment on their computers.

As early as April 1976, Rediffusion Engineering Limited in England installed a color TV cable (fiber optic) over a 1.5-km stretch in Hastings, Sussex. Significant advances have and are being made in fiber-optic technology in all electronics fields.

NEW INNOVATIONS IN FIBER OPTICS

Vehicles used to lift personnel into high-tension lines may use fiber optics to reduce the chance of power spikes and the risk of injuring personnel by moving them into high-tension lines. Information could be taken to miners by fiber optics and therefore reduce the chance of spark generation.

OPERATION OF A BASIC FIBER-OPTIC SYSTEM

A typical fiber-optic system is illustrated in figure 2–1. The operation of this simple system is defined in simple terms here and will be described later in detail. A system consists of the following:

1. A transmitter accepts an electrical signal and converts it to a current to drive the light source.
2. The light source launches the optical signal into the fiber.
3. The optical fiber provides a path for the optical signal.
4. A light detector detects and converts the optical signal to an electrical signal.
5. A receiver produces low noise and large voltage gain from the power detector signal.
6. Various connectors and splices interface the system.

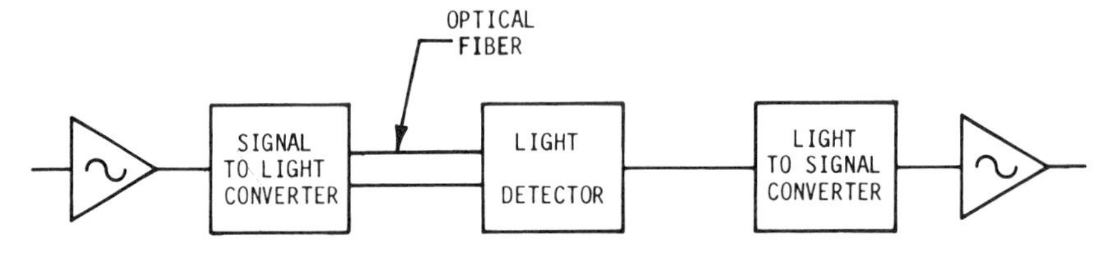

Figure 2–1 Typical Fiber-Optic System

OPTICAL FIBERS

Optical fibers are transparent, dielectric cylinders surrounded by a second transparent dielectric cylinder. The fibers are light waveguides used to transmit energy at optical wavelengths. The light is transported by a series of reflections from wall to wall of an interface between a core (inner cylinder) and its cladding (outer cylinder). The reflections are made possible by a high refractive index of the core material and a lower refractive index of the cladding material. Refractive index is a measure of the fiber's optical density. The abrupt differences in the refractive indexes causes the light wave to bounce from the core/cladding interface back through the core to its opposite wall (interface). Thus the light is transported from a light source to a light detector on the opposite end of the fiber.

Fiber Generalities

The discussion of how light is launched into a fiber and arrives on the other end of the fiber is fairly easy to understand. The details involved in transmission and the mechanics of coupling, however, are manifold. Let us first discuss in lay terms how transmission takes place.

Light is propagated through optical fibers in a series of reflections from one side wall to opposite wall. The acceptance angle *i* must be small enough so that all of the signal is reflected. In figure 2–2, a fiber core is covered by a layer called the *cladding*. The core has a refractive index higher than that of the cladding. Index of refraction is determined by the manufacturer when the fiber is produced. When the light enters the optic fiber, it reflects from wall to wall of the cladding back into the core. The fiber illustrated is a step-index fiber. That is, there is an abrupt refractive index change between the core and the cladding. In order that the light

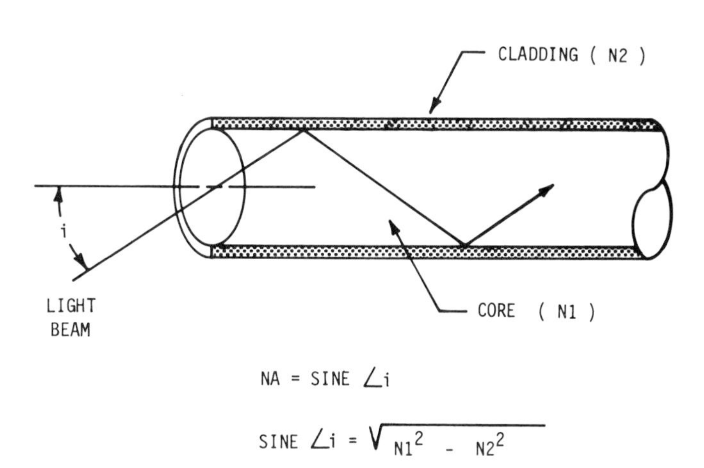

Figure 2–2 Basic Optical Fiber

beam may bounce freely from wall to wall without loss, the cladding was added. The cladding acts as a mirror to reflect without power loss. A second use of the cladding is to prevent light from escaping to nearby fibers in the event of bundling several together. The cladding also contributes to the strength of the fiber.

The acceptance angle i is one-half the total cone of acceptance. The cone of acceptance is the light-gathering area at the fiber input. The mechanics of transmission angles are handled in detail in a later chapter.

Fiber Properties

There are six major properties involved when choosing the correct fiber for an optical system. These are:

1. *Numerical aperture (NA).* Numerical aperture is the mathematical sine of the angle of acceptance. Half the cone of acceptance is the angle i in figure 2–2.

$$\mathrm{NA} = \sin \angle i$$
$$\mathrm{NA} = \sqrt{n_1^2 - n_2^2} = \sin \angle i$$

where n_1 = refractive index of the core
n_2 = refractive index of the cladding
$\angle i$ = angle of acceptance

NA is a trigonometric function which represents the sine of the angle i. If the NA increases, the angle i must have also increased and the fiber sees more light. This may seem like a good thing. A problem enters when there is a greater spread in propagated light rays. The increase in ray velocity limits the bandwidth. The effect is called *intermodal dispersion.* The NA of a fiber can never be greater than 1.0. This is mathematical maximum sine value. Most NA values are low, between 0.20 and 0.60, and seldom near the maximum sine value. This is a range and must not be judged. For specific values, the manufacturers' data sheets must be consulted.

2. *Dispersion.* The second property to contend with is *dispersion.* Dispersion is the spreading or widening of light rays. In fiber-optic systems, dispersion is either intermodal or intramodal. The *inter-*

modal dispersion (also called *multimode*) is the propagation of rays of the same wavelength along different paths through a fiber. This results in the wavelengths arriving at the opposite end of the fiber at different time periods. The second dispersion, *intramodal,* is determined by three separate factors: material dispersion, waveguide dispersion, and cross-product dispersion. *Material dispersion* is due to variations of the index of refraction of the core and the cladding materials. *Waveguide dispersion* is a phenomenon caused by the bandwidth of the signal and the waveguide configuration. *Cross-product dispersion* is leakage of optical energy from one material, fiber, cable, or connector to other materials. Wavelength and cross-product dispersions are usually small and can be ignored. Dispersion is a function of the refractive index of the material the fiber is made of and the wavelength or mode of the light traveling through the fiber. It is measured in nanoseconds per kilometer (ns/km).

3. *Attenuation.* Attenuation is the loss or reduction in amplitude of transmitted energy. Intrinsic losses are due to light scattering of transmitted energy. Losses of light due to the differences of refractive indexes of two mediums at their interface are called *Fresnel losses.* Intrinsic losses are due to light scattering within the fiber. The most prevalent type of light scattering is Rayleigh. *Rayleigh scattering* is power loss due to the molecular structure of the fiber material. Other power loss may be due to imperfections or bubbles in the fiber material. Man-made power losses from scratches or dirt are common.
4. *Bandwidth parameters.* Attenuation curves are always provided by manufacturers of fibers to allow the designer to choose the correct fiber for a particular use. Curves or equivalent tabulated data are expressed in decibels per kilometer (dB/km). The curve in figure 2–3 plots wavelength against decibels. This provides the designer with the fiber's transmission qualities. These parameters are developed during manufacture.
5. *Rise time.* The rise time is a parameter that tells the designer that selected parts will operate at speed required. Rise time identifies the fiber's dispersion properties. Rise time is determined by intermodal (multimode) dispersion and/or intramodal (material) dispersion. To find the total rise time of a system, you simply add the rise time required for each time-critical component, then add some

tolerance. In their *Technical Note R-5,* ITT Electro-Optical Products Division uses this simple calculation. The overall system rise time is 1.1 times the root sum square of all system constituents:

$$T_{\text{system}} = 1.1\sqrt{T_I^2 + T_M^2 + T_{\text{etc.}}^2}$$

where T_I = intermodal rise time

T_M = material intramodal rise time

6. *Fiber strength.* Another fiber property is tensile strength, which is governed by manufacturing intricacies. These techniques eliminate flaws and microcracks in the fiber. Flaw-free fiber cores, clad-

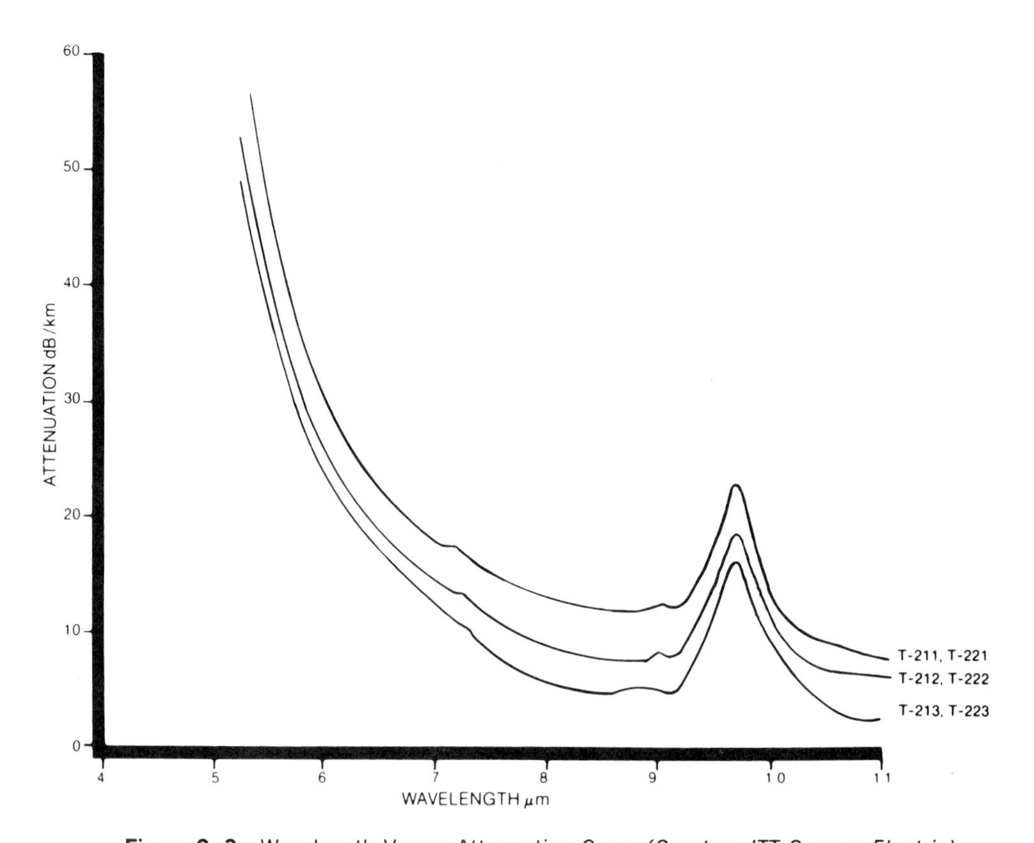

Figure 2–3 Wavelength Versus Attenuation Curve *(Courtesy ITT Cannon Electric)*

dings, and surfaces are the ultimate attempt in the fabrication of excellent fibers.

It must be noted that these six parameters (numerical aperture, dispersion, optical attenuation, bandwidth and frequency curves, rise time, and fiber strength) are necessary requirements. Additional considerations during design may be size, shape, and bend radius, among others.

FIBER TYPES AND TRANSMISSION MODES

Fibers are constructed to fill the need of a particular requirement. The major need is in the area of bandwidth, that is, low, medium, high, or ultrahigh. In addition, a fiber may be constructed for its extremely low attenuation per kilometer. In some cases tensile strength is the most desired characteristic. Two operating classes of fibers are utilized, single mode and multimode. *Single-mode fibers* are fibers that carry one mode. Single-mode fibers must be designed to accommodate a specific wavelength; otherwise, large attenuation will result. An attractive feature of the single-mode fiber is that it is not sensitive to microbending. These are losses induced by local lateral displacements of the fiber. *Multimode fibers* allow intermodally dispersed wavelengths. That is, the propagation of rays of the same wavelength follow different paths through the fiber, causing different arrival times.

There are two types of optical fibers, step index and graded index. Cross-sectional views along with the typical light-ray path of these fiber types are illustrated in figure 2–4. The *step-index fiber* has an abrupt refractive index between the core and the cladding. The *graded-index fiber* has a variation of refractive index between the core and the cladding. This minimizes dispersion and allows a greater bandwidth of information to be transmitted.

Fibers are constructed from plastic or glass or a combination of the two: for example, glass core/glass cladding, plastic core/plastic cladding, or glass core/plastic cladding. The choice of a fiber type has to do with the quality, the parameters required, and economics. Can the system live with dispersion or must we use a better fiber at a higher price? In real life there are trade-offs in design. You cannot always receive lower cost without lower performance. A lower attenuation per kilometer may also mean a very high price tag.

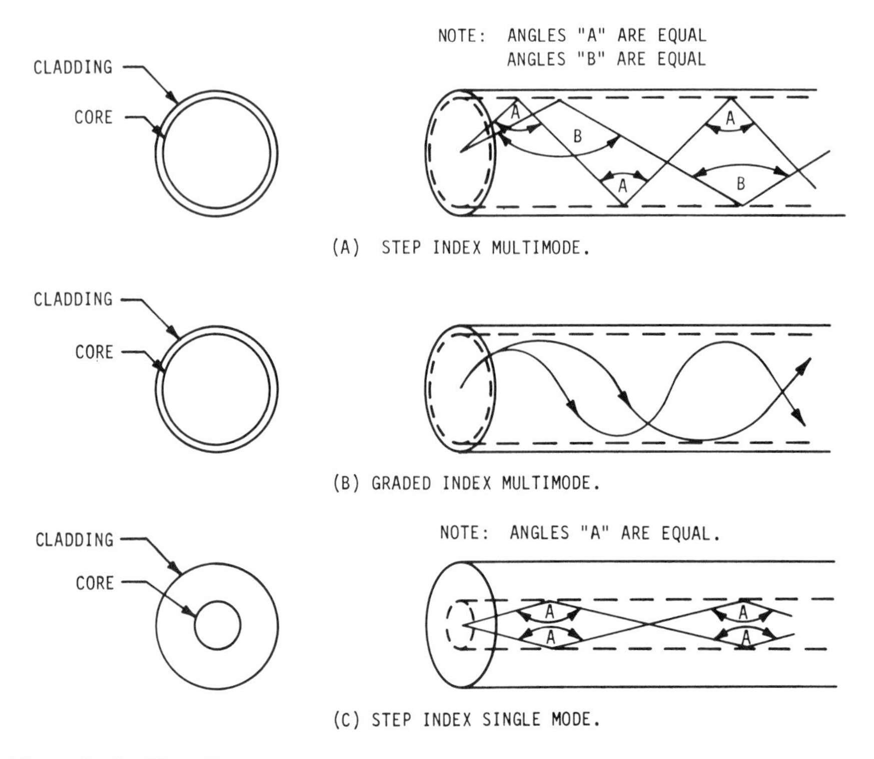

Figure 2–4 Fiber Types

CABLE TYPES

Cables of single or multiple fiber are manufactured usually for protection of the fiber. Fibers are encased within the cable using protective jackets (usually polyurethane). Strength members surround the fibers. Cables are constructed to feature light weight, high flexibility, resistance to kinks, strength, and resistance to crushing. Some tests imposed on cables in the development stage are impact, bend, twist, fatigue under load, high and low temperature, and even storage losses.

A BASIC FIBER-OPTIC SYSTEM

A simple fiber-optic system is called a *transmission link*. It consists of a transmitter with a light source, a length of fiber, and a receiver with a light detector. The operation of a system is as follows. A digital or analog signal is applied to a transmitter. Within the transmitter, the input signals are converted from electrical to optical energy by modulating an optical light source, normally achieved by varying current. The modulated light is launched into a length of fiber, where it reflects from wall to wall through the fiber core. At the opposite end of the fiber, a detector accepts the light signal and converts it back to an electrical signal. The electrical signal may now be conditioned to perform its original work. Consider now a brief discussion of the several components that make up a fiber-optic system.

Electrical Signal Transmitter (Driver)

The purpose of the transmitter (driver) is to change the electrical signal into the required current to drive a low-impedance light source. The electrical inputs are either digital or analog. The choice of the converter should depend on the current requirement of the light source. If the signal is digital, the transmitter (driver) should consist of a high-speed pulser to turn the light source on and off. If the signal is analog, the transmitter (driver) should be able to supply current to the light source to transmit the positive and negative alternations of the signal.

A typical analog driver is illustrated in figure 2–5. R_2 and R_3 provide a voltage divider for the input signal. The potentiometer P_1 and resistor R_1 serve to set the operating point so that the positive and negative swings of the analog input signal produce only a positive output. The output current never changes polarity, only amplitude. R_{FB} is the feedback resistance. The light-emitting diode (LED) transmits light as current varies. Resistor R_L provides current limiting and load.

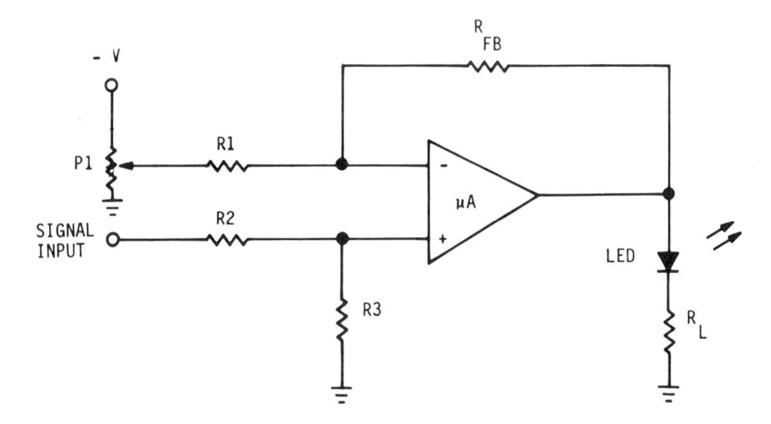

Figure 2–5 Typical Analog Driver

In figure 2–6A, a very basic LED driver consists of an inverter with an LED tied to its output. The resistor R_L serves as a current limiter. Signal pulses at the input direct pulse current to flow through the LED causing it to radiate.

In figure 2–6B, a variable (analog) signal input is directed through a pulse modulator. The modulator can be one of three types: PRM—pulse-rate modulation, PPM—pulse-position modulation, or PWM—pulse-width modulation. Modulated output directs pulse current to flow through the LED, causing it to radiate.

The samples shown in figures 2–5 and 2–6 are extremely simplified.

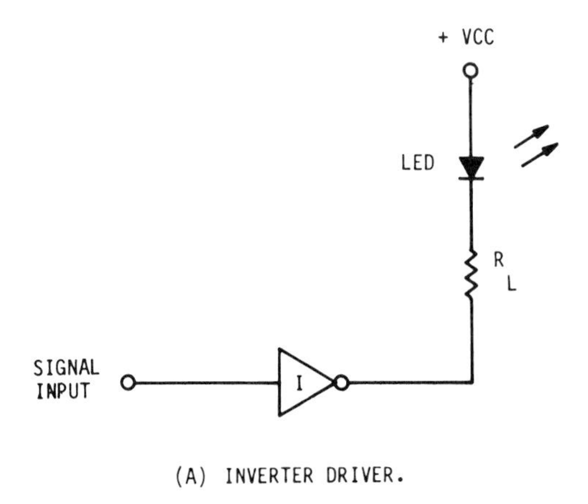

(A) INVERTER DRIVER.

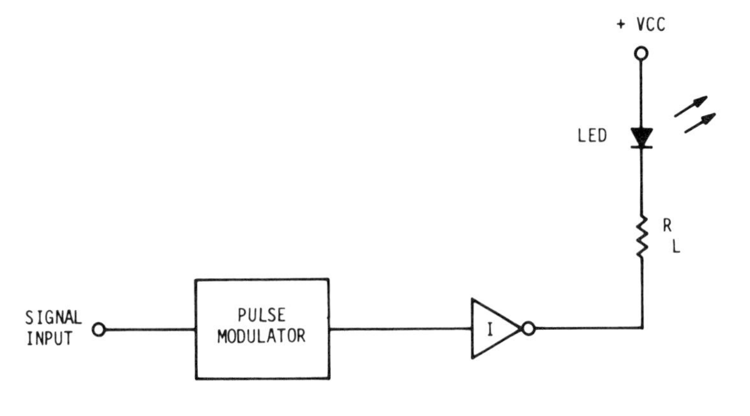

(B) PULSE MODULATED DRIVER.

Figure 2–6 Basic LED Drivers

Light Source

The purpose of the light source is to launch a light signal into the optical fiber at an angle that provides maximum signal transfer. There are two basic light sources used in fiber-optic electronics. These are the light-emitting diode (LED) and the injection laser diode (ILD). Both of these units provide small size, brightness, low drive voltage, and are able to emit signals at desired wavelengths. Each has characteristics that make them desirable or undesirable for a particular application. The LED has a longer life span, greater stability, wider temperature range, and much lower cost. The ILD is capable of producing as much as 10 dB more power output than the LED. It can launch the light signal at a much narrower numerical aperture (NA), and therefore can couple more power through the optic fiber than the LED. The disadvantage of using an ILD is that its current range is extremely restricted. Since some system operations vary greatly, the ILD must have compensation devices added to the electronics. This may make the cost prohibitive.

Light Detector

The light detector accepts a light signal from the optical fiber and converts it into an electrical current. There are three types of detectors in use: the phototransistor, the PIN diode, and the APD. The *phototransistor* is inexpensive but has slow rise times and a limited bandwidth capability. Therefore, it is seldom used. The *PIN diode* (contains positive, intrinsic, and negative solid-state layers in its construction) exhibits fast rise time and acceptable bandwidth parameters. It is reasonably priced. The *APD diode* (avalanche photodiode) exhibits fast rise time and acceptable bandwidth parameters. The APD is more expensive than the PIN diode since it provides greater receiver sensitivity. The APD also requires an auxiliary power supply.

Receiver

The function of the receiver is to accept low-level power from the detector and convert it into a high-voltage output. There are at least two methods of accomplishing this. In figure 2–7A, detector current produces a voltage drop across a load resistance. The voltage drop is directed into an ampli-

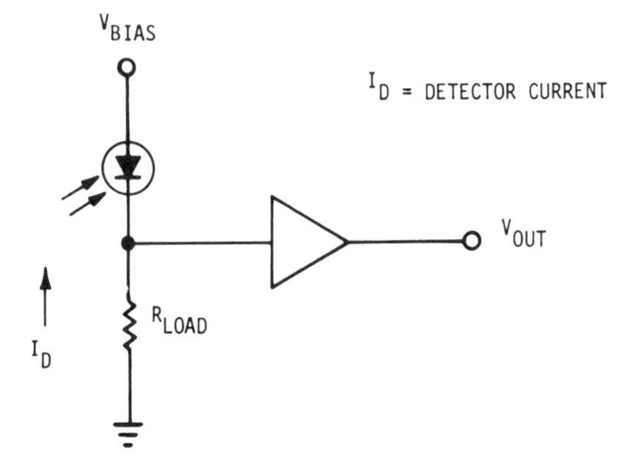

(A) DETECTOR CURRENT CAUSING VOLTAGE DROP ACROSS A LOAD RESISTOR.

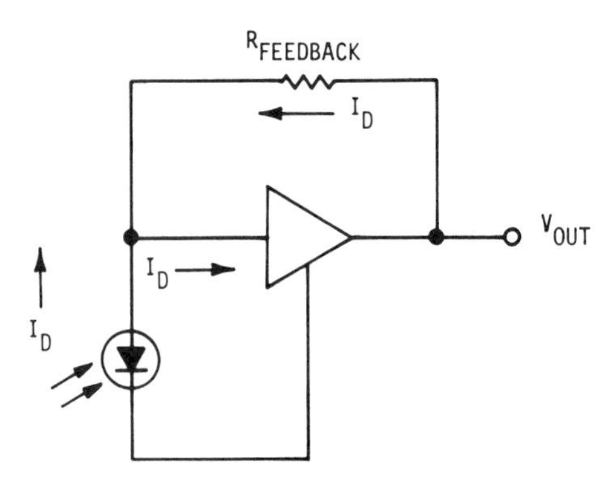

(B) AMPLIFIER DRIVES DETECTOR CURRENT THRU FEEDBACK RESISTOR.

Figure 2–7 Receiver Action

fier. An output voltage representative of the transmitted signal is the result. In figure 2–7B, the operational amplifier output voltage is the effect of the amplifier driving detector current through the feedback resistance. Again the output voltage is representative of the transmitted signal.

Other electronics may be added to the circuitry to maintain correct response. A gain control may be used on the front end to vary the impedance of the receiver. The operational amplifier is used as a current-to-voltage converter.

Since signal inputs are generally weak, shielding and power-supply decoupling are a requirement to achieve sensitivity. Sensitivity is set by input noise. Noise, of course, leads to errors in digital systems and restricts signal-to-noise ratio (SNR) in analog systems.

3

Fiber-Optic Geometry

You may recall from discussion in chapter 2 that light propagates along an optical fiber by a process called reflection. This is, of course, if all the criteria for this reflection are acceptable. The fiber core must be reasonably pure material. The core must be surrounded by a second material which has a refractive index lower than the core. The outer insulation is called a *cladding*. Light may be propagated by an injection source. The launch is made at specific angles to ensure maximum transfer of power. Transmission is much the same as radio electromagnetic wave transmission. The exception is that light is being transmitted. The light is received at the opposite end of the fiber and converted to an electrical signal. The electrical signal may be utilized to perform a typical function, either digital or analog. Transmission and power are dependent on many specifications and parameters. Perhaps the most important of these are the transmission angles.

ANOTHER LOOK AT LIGHT ANGLES

You may recall from the previous chapters that incoming rays of light incident to a surface either reflect or refract from that surface. That is, they bounce off the surface or penetrate into the new medium. The angles that are involved are the angles of reflection and the angles of refraction. These angles are dependent on the material surface structure and the index of refraction of the two media. The optical fiber ends must be considered smooth. The index of refraction is a ratio of the speed of light in a vacuum to the speed of light in the optical fiber. These indexes are available from the fiber manufacturers and are listed in physics books.

Reflection

When light beams or rays strike a surface, they reflect from that surface either in a diffuse manner or a specular manner or in some cases both. The amount of reflection is largely due to the type of material on the surface of the reflector. (See figure 3–1.)

Diffuse Reflection

Diffuse reflectors scatter the light outward in several beams, depending, of course, on the surface material and the amount of light the material absorbs. Virtually all material absorbs some light. The principle of diffuse reflection is illustrated in figure 3–1A. A smooth surface such as a freshly painted house would diffuse light in rather equal and alike reflected rays. Another surface, such as a rock-covered roof, would cause light rays to be reflected in scattered, nondescript lines.

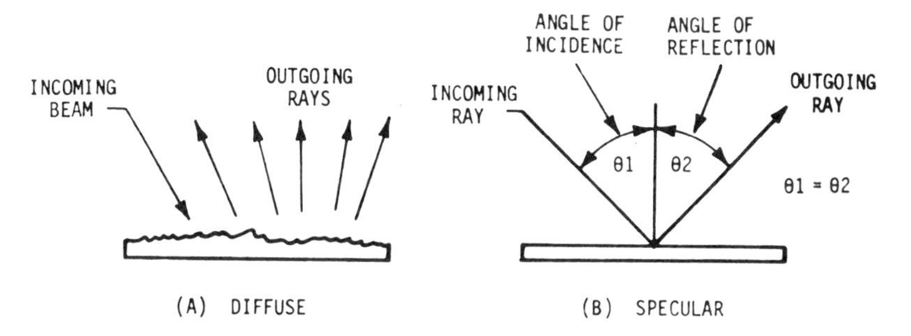

Figure 3–1 Reflection

Surface or Specular Reflection

Whenever a beam of light from one medium such as air, strikes a second mirrorlike medium such as glass, part of the beam is reflected and the other part refracted. (See figure 3–1B.) The angle that the incident rays strike the second medium is called the *angle of incidence*. This angle is the angle made by the incident beam and a line normal (perpendicular) to the boundary of the two media. Part of the incident beam is reflected at an angle called the *angle of reflection*. This angle is made by the reflected beam and a line normal (perpendicular) to the boundary of the two media:

Angle of incidence = Angle of reflection

Refraction (See figure 3–2)

When a beam of light is propagated on a surface that separates two media such as air and water, it is reflected back into the medium from which it came. Part of the light is transmitted into the second medium. Let us first consider the reflected rays. The direction of the propagation is called the *direction of the incident*. A vertical line perpendicular (90°) to the surface is called the normal. The direction of reflection is on the opposite side of the normal from the propagated beam. These three directions (incident, normal, and reflection) are coplaner, that is, in the same plane. The angle made between the direction of incidence of the light beam and the normal is the angle of incidence. The angle between the normal and the direction of reflection is the angle of reflection.

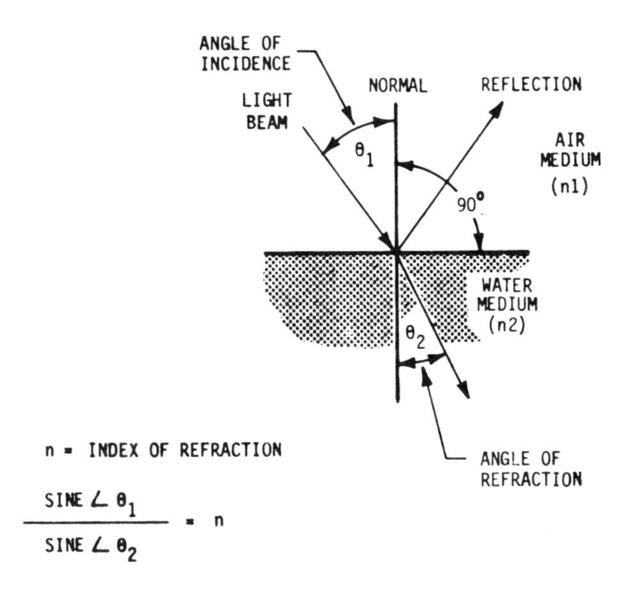

n = INDEX OF REFRACTION

$$\frac{\text{SINE } \angle \theta_1}{\text{SINE } \angle \theta_2} = n$$

Figure 3–2 Surface Refraction

Now let us consider a second set of angles and direction. Part of the light beam is transmitted into the second medium. These portions of the light beam do not follow the direction of incidence but are bent (refracted) at the surface. The direction of propagation, the normal, and the direction of refraction (bending) are also coplaner. The angle between the normal and the direction of refraction is the angle of refraction. The relationship between the angle of incidence and the angle of refraction is a constant ratio. This is a ratio between the refractive indexes of the two materials and mathematically is developed in the following manner:

$$n \text{ (refraction constant)} = \frac{\sin \angle \theta_1}{\sin \angle \theta_2}$$

Refraction and the refractive index are utilized in the design of optical equipment. The index for optical glasses varies from 1.26 to 1.96. Air has a refractive index of 1.0, whereas a diamond has an index near 2.4. The index is a measure of optical density and may be found in most general physics texts.

The angle of refraction is extremely important in transmission of light beams into optical fibers. The critical angle at which the light beams are reflected within the medium of greater density is determined by its refractive index. This critical angle is used in design of fiber-optic systems. The relationship between the angle of incidence and the angle of refraction is defined by Snell's law, which states:

$$n_2 \sin \angle \theta_2 = n_1 \sin \angle \theta_1$$

or

$$\sin \angle \theta_2 = \frac{n_1}{n_2} \sin \angle \theta_1$$

where n_1 and n_2 represent refractive indexes of the two media as illustrated in figure 3–2. The angle θ_1 is the angle of incidence. The angle θ_2 is the angle of refraction.

SNELL'S LAW IN RELATION TO OPTICAL FIBERS

The physics of refraction is explained in terms of Snell's law (see figure 3–3) when applied to a optical fiber end. Rays intersecting a plane surface (fiber end) are refracted in the following relationship. The index of refraction of the air times the sine of the angle of incidence is equal to the index of refraction of the fiber times the sine of the angle of refraction:

$$n_a \sin \angle a = n_r \sin \angle r$$

where n_a = index of refraction of air

n_r = index of refraction of fiber

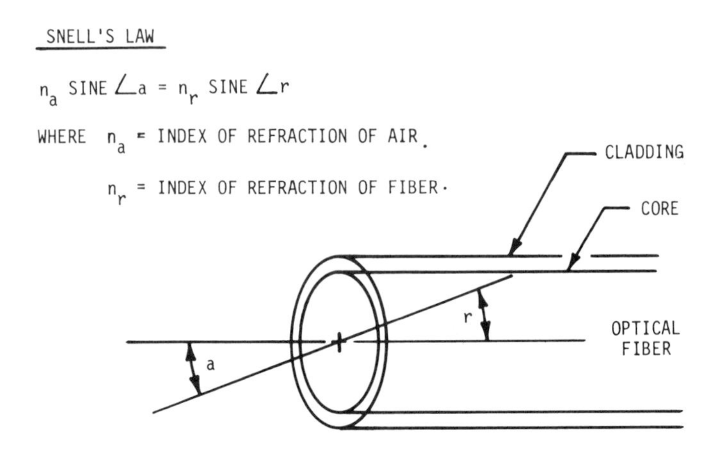

Figure 3–3 Snell's Law

TRANSMISSION ANGLES

Angles of incidence, reflection, and refraction play a big part in the explanation and understanding of optical fiber transmission angles. Let's consider some of these angles in their optical fiber realm.

Cone of Acceptance (See figure 3–4.)

The cone of acceptance is the area in front of a fiber face that determines the angle of light waves that will be accepted into the fiber. A fiber may accept and propagate many rays of light which are less than the angle presented by the cone of acceptance. Light enters at an infinite number of angles within this cone of acceptance. The angle at which the rays are accepted is called the *angle of acceptance*.

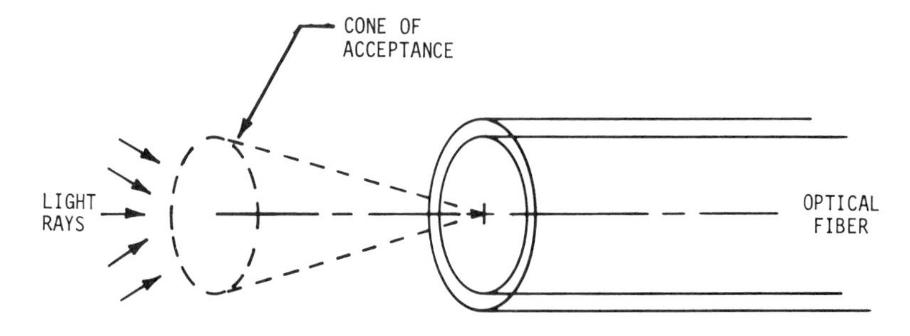

Figure 3–4 Cone of Acceptance

Angle of Acceptance (See Figure 3–5)

The angle of acceptance is an angle that represents the half-angle of the cone of acceptance. The mathematical sine of the angle of acceptance is called the numerical aperture (NA). Numerical aperture may be calculated by several means:

$$\mathrm{NA} = \sin \angle a$$

or

$$\mathrm{NA} = \sqrt{n_1^2 - n_2^2}$$

or

$$\mathrm{NA} = (n_1^2 - n_2^2)^{1/2}$$

where n_1 = index of refraction of the core
n_2 = index of refraction of the cladding
$\angle a$ = acceptance cone half-angles

The reader will note that the angle a ($\angle a$) is usually designated $\angle \theta$ in engineering circles. It was felt, though, for explanation purposes, that we would use the first letter of the angle title. The numerical aperture (NA) can also be construed as the maximum acceptance angle, which, in turn, defines the light-gathering power of the fiber, that is, the measure of the fiber's ability to accept light waves and reflect them through the fiber.

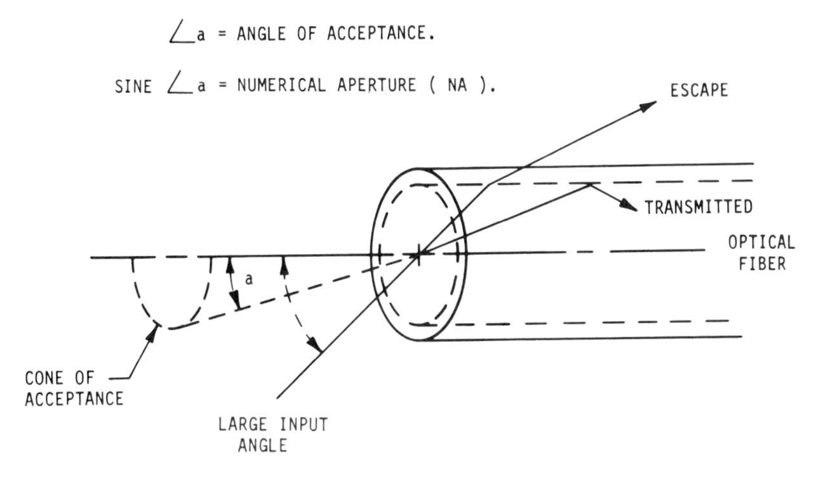

Figure 3–5 Angle of Acceptance

Angle of Refraction (See Figure 3–6)

The angle of refraction is the angle at which the rays are refracted from the cone of acceptance into the core (see figure 3–6). Light waves are propagated from air at $\angle a$. These waves are refracted at the surface of the optic fiber into the core at $\angle r$. Therefore, the angle of incidence and the angle of refraction are related, using Snell's law. Calculations using Snell's law are as follows:

$$n_a \sin \angle a = n_r \sin \angle r$$

Therefore,

$$\sin \angle a = \frac{n_r}{n_a} \sin \angle r$$

Remember that the numerical aperture NA $= \sin \angle a$, which represents the maximum acceptance angle. Rays entering the fiber at a greater angle than $\angle a$ are lost and will not reflect. Rays entering the fiber at a smaller angle than $\angle a$ are totally reflected.

Angle of Reflection

Propagated waves at the angle of refraction ($\angle r$) strike the interface of the core and cladding. The angle of refraction at this point is actually an angle of incidence and the waves are reflected from the core/cladding interface at the same angle at which they strike the interface. In figure 3–6,

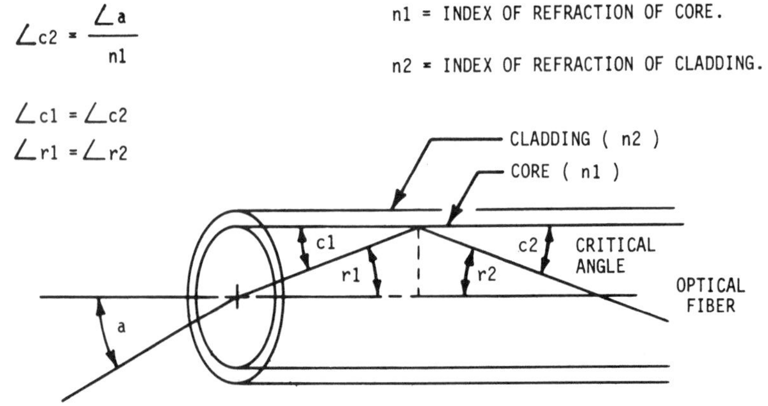

$$\angle c2 = \frac{\angle a}{n1}$$

$\angle c1 = \angle c2$

$\angle r1 = \angle r2$

n1 = INDEX OF REFRACTION OF CORE.

n2 = INDEX OF REFRACTION OF CLADDING.

CLADDING (n2)
CORE (n1)
c1
r1
c2
r2
CRITICAL ANGLE
OPTICAL FIBER
a

Figure 3–6 Transmission Angles

the angles r_1 and r_2 are equal and the angles c_1 and c_2 are also equal. The angles c_1 and c_2 are called internal *critical angles*. Any mode that strikes the core/cladding interface at an angle greater than c will not be reflected but lost in the cladding. Modes that strike the interface at a smaller angle than c will be reflected through the fiber. The critical angles are also functions of the core and cladding indexes of refraction. The internal critical angle c_2 is related to the fiber indexes of refraction in the following manner.

$$\angle c_2 = \frac{\arcsin \sqrt{n_1^2 - n_2^2}}{n_1}$$

$$\angle c_2 \simeq \frac{a}{n_1}$$

Graded-Index Fiber Angles (See Figure 3–7)

Rays reflected at higher angles (do not launch from at center of the core) have to travel a greater distance through the optical fiber. The time for propagation is therefore longer. This causes an effect called pulse *broadening* or *dispersion*. A graded-index fiber is used to reduce dispersion within the fibers (see figure 3–7). The numerical aperture (NA) is calculated in the same manner as with the step-index fiber. The graded-index core has an index of refraction that decreases with radial distance. The light waves travel faster near the cladding and slower near the core center. These phenomena allow the speed of all waves at all angles to be approximately the same. In actual practice there are three ray paths. Rays in the center of the fiber travel in high-index-of-refraction (slow) material but travel the shortest path. Rays taking a sinusoidal path through the fiber cover a greater distance than the center rays, but travel in low-index-of-refraction (fast) material, part of the time. Finally, rays traveling in a helical path cover a

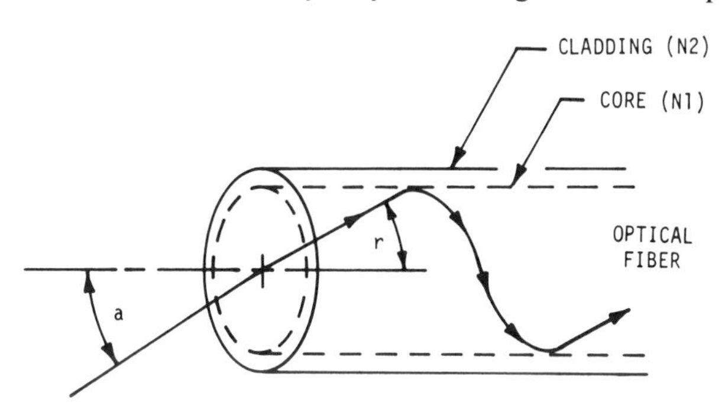

Figure 3–7 Graded-Index Fiber Ray Paths

greater distance, but travel in fast material (low index of refraction). This is an improvement over the step-index fiber in that larger angles in the step index take longer to reach the receiving end of the fiber.

Light Sourcing (See Figure 3–8)

A point source is theoretically a dimensionless (no length, width, or height) point in space from which light is propagated in all directions away from the source. In reality this cannot be, for any point or light source must have a finite size. Point sources can, however, appear to be a defined point such as a star that is light-years away. An extended source is depicted as a point in space that is illuminated by light radiated from several different directions.

The light rays appear more nearly parallel the farther you move away from a point source. From an extreme distance, the extended source appears to be a point source.

The extended source of light is used in fiber optics. The intensity of this source is the amount of radiant flux diverging from the source. This flux is called *exitance*.

An area source whose angular distribution is a circle is called a *Lambertian source*. (See figure 3–9.) The flat LED (light-emitting diode) pro-

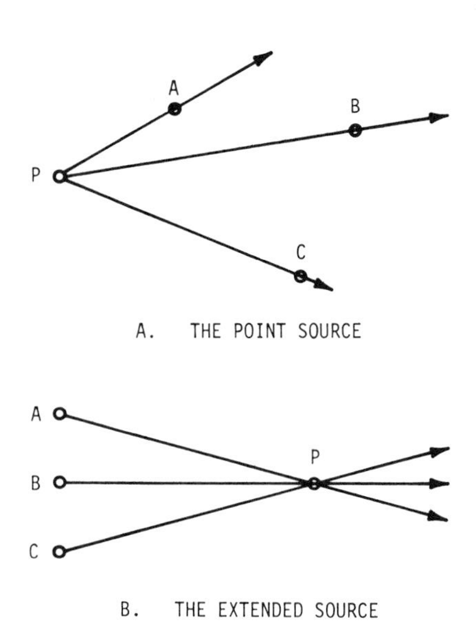

Figure 3–8 Basic Light Sources

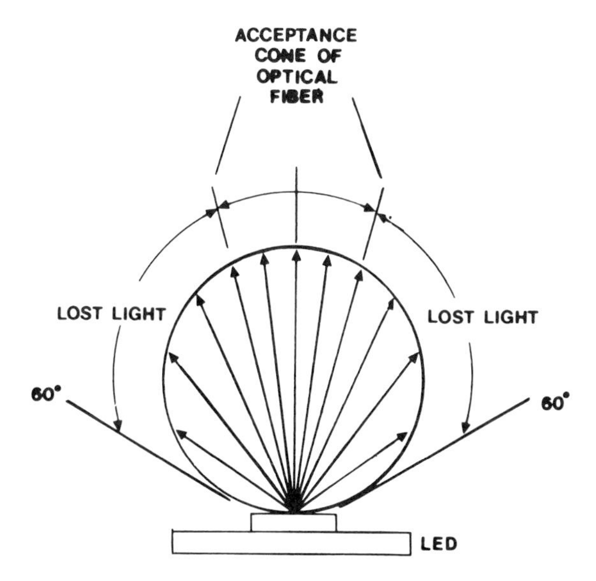

Figure 3–9 The LED as a Lambertian Source of Light

vides an excellent approximation of the Lambertian source. When coupling the LED to the end of an optical fiber, the radiation pattern must be considered. The pattern is matched to the acceptance cone of the fiber so that the maximum light signals are launched into the fiber.

4

Alignment Philosophy

Simply because of its size, the coupling of fibers is an intricate, demanding task. Fiber ends must be terminated. Special attention must be made to align source light to the fiber and to the detector on the receiving end. Fiber-to-fiber splices are critical. The loss of decibels of power seems insignificant. If you multiply those few decibels by several joints throughout a system, the power loss may become unacceptable.

CAUSES OF COUPLING POWER LOSS

Before we develop procedures defining alignment of optical fibers, it seems appropriate to discuss some of the causes of power loss in coupling. Power losses are rated in decibels (dB) and are additive regardless of their reason for existence. Some of these losses are as follows:

1. Heat is always a problem in dealing with any engineering effort. Optical fibers must be isolated with heat sinks tied to receptacles, mountings, connectors, and certain splices.
2. Fiber isolation, when in a cable or bundle, may cause some power loss. Sealing off adjacent light sources is necessary. Opaque fiber coverings and sheathing are often added to protect against light sources and the environment.
3. If cable ends are allowed to vibrate, the vibration may actually cause modulation of the light rays.
4. Sand or dust on the critical mirror-finished fiber ends may act as

an abrasive. Pitted finish may reflect light rays in a diffuse rather than a specular manner.

5. Water or humidity at the connection may actually improve the coupling. Water has an index of refraction much nearer the fiber core than air.

These are just some of the reasons for power loss. The most prevalent of coupling power loss is due to misalignment of the fiber. There are four major forms of fiber misalignment. These are core irregularity, lateral misalignment, gap loss, and angular loss. A description of each is provided in the next several paragraphs.

Core Irregularity

During manufacture, the core of the fiber may be placed off center within the fiber. In extremely long fiber lengths there may be short stretches of core that are off center. In figure 4–1, cores are shown off center. The amount is exaggerated for understanding.

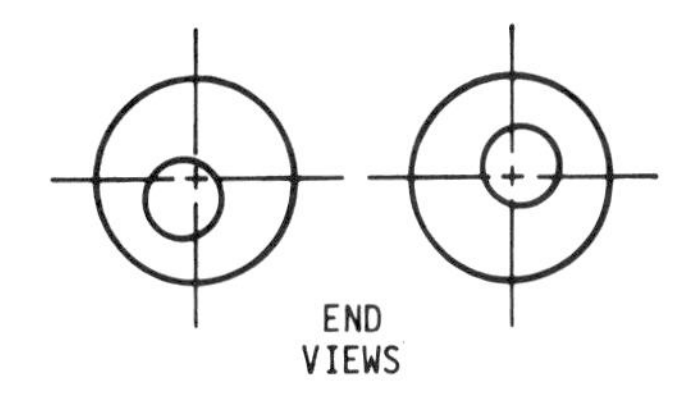

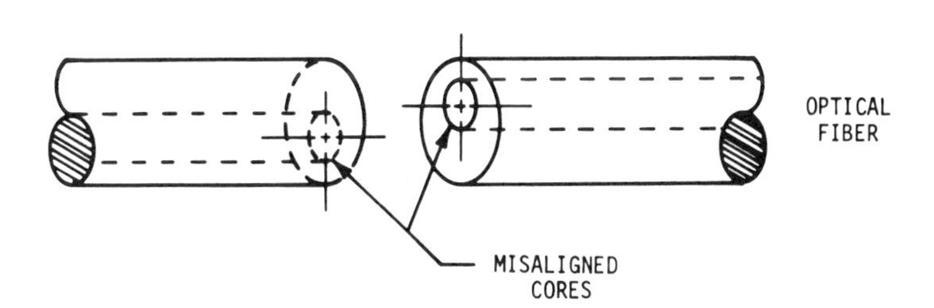

NOTE: EXAGGERATED CORE IRREGULARITY.

Figure 4–1 Core Irregularity

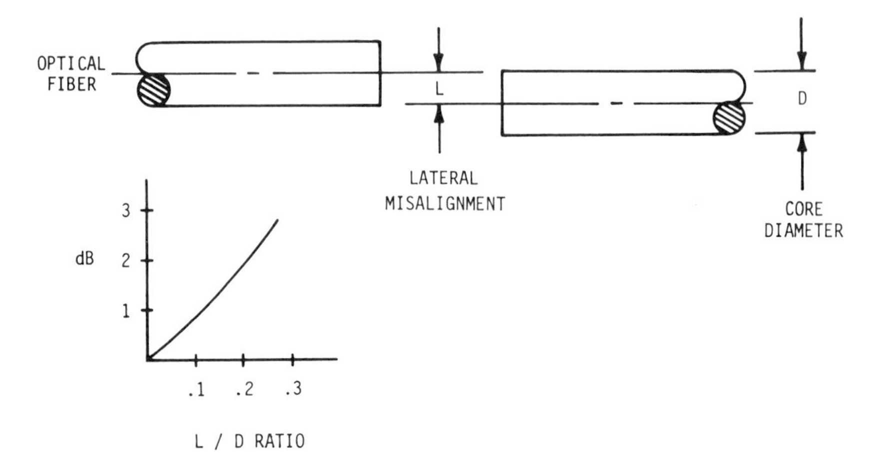

Figure 4–2 Lateral Misalignment

Lateral Misalignment

The centerlines of cores may be misaligned. Axial misalignment, as this is often called, is the lateral displacement between centerlines of the cores in the two mating fibers (see figure 4–2). Power loss in decibels is plotted as an *L*/*D* ratio, where *L* is the lateral displacement and *D* is the core diameter. Centerlines should be between 0.0001 and 0.0002 in. as specified, depending on fiber diameter.

Gap Loss

Gap loss is the separation of the two mirror-finished fiber ends (see figure 4–3). Gap loss is expressed in decibels and is determined by the NA of the fiber. Lower value NAs have less gap loss than higher NAs. Gap loss is plotted as a *G*/*D* ratio against decibels, where *G* is the gap and *D* is the core diameter. Typical allowable gaps are 0.0002 or 0.0003 in. or as specified.

Angular Loss

Angular loss is caused by cutting the fiber ends at an angle or by preparing the faces of the fibers so that they are not symmetric (see figure 4–4). Again, the power loss is expressed in decibels. Angular losses are plotted

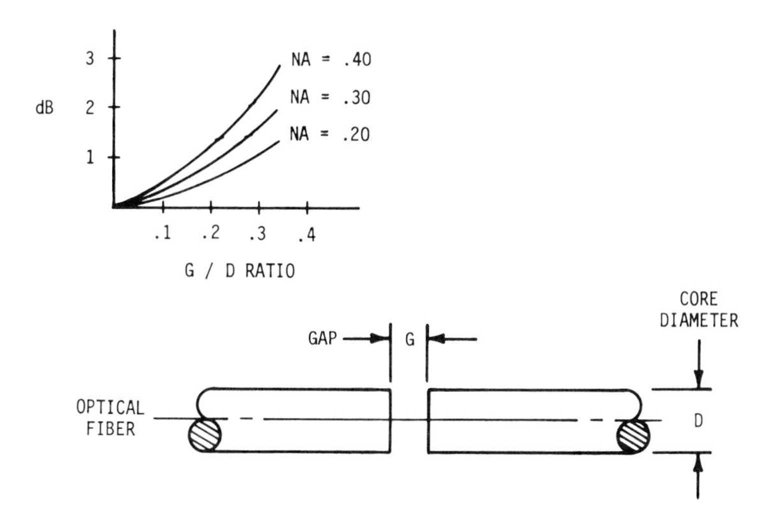

Figure 4–3 Gap Loss

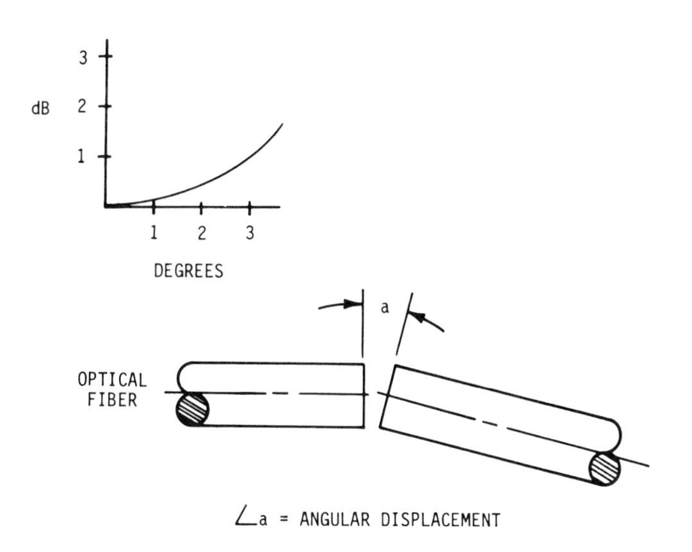

Figure 4–4 Angular Loss

as angles in degrees against decibels, where the angle is the shape between the two mirror fiber end faces. A typical allowable angle is less than ¼°.

Loss in Transmitting from a Large Fiber to a Smaller One

When two fibers of different diameter are mated together, a loss of optical energy will result. Calculation of this loss is made in the following manner:

$$dB = -10 \log \left(\frac{D_r}{D_t}\right)$$

where D_r = diameter of the receiving fiber
D_t = diameter of the transmitting fiber

Losses in Transmission Due to Changes in Numerical Aperture (NA)

When two fibers of different numerical aperture (NA) are joined together, the loss of optical energy may be calculated in the following manner:

$$dB = -10 \log \left(\frac{NA_r}{NA_t}\right)$$

where NA_r = numerical aperture of receiving fiber
NA_t = numerical aperture of transmitting fiber

Coupling Attenuation

Attenuation in coupling is established by two general criteria. These are intrinsic losses and extrinsic losses. Intrinsic losses are due to factors involved in the fiber and its manufacture. Intrinsic losses have nothing to do with the connector or its installation. Several producers of intrinsic loss are thus:

1. *Core diameter variations.* The core diameter variations are generally minute but can cause loss.
2. *Fresnel reflection.* Fresnel loss is due to the difference in the index of refraction of the air and the fiber at the fiber end.

3. *Transmission loss.* Losses due to the light waves lost in the cladding and not reflected from its walls.
4. *Absorbtion loss.* Loss due to nontransparent core material.

Extrinsic losses are usually attributed to the connector hardware and man-made misalignment. Connectors and the details of misalignment and/or alignment are discussed under separate headings.

ALIGNMENT MECHANICS

The mechanics of alignment are an exacting task in that perfect or at least optimum alignment is necessary to avoid loss. This philosophy must extend to the retaining of optimum alignment in every instance that a connection is made or remade. Furthermore, alignment must be of a method that can be accomplished in the field with a minimum of special equipment. Alignment should also be within the technical capabilities of available personnel who are already maintenance technicians.

Under the next several headings, we shall study several of the mechanical concepts used in the fiber industry today. Many of these were invented and patented by individuals and then were developed by major corporations for use in coupling devices.

Opposed-Lens Alignment Concept

A pair of highly polished fiber ends are inserted into opposing cavities within a transparent medium. An objective lens and an immersion lens are provided by the opposed cavities in the transparent medium. The viscous fluid has an index of refraction that allows alignment of the opposing fibers. In figure 4–5, light rays are shown refracted from the viscous indices, through the transparent medium aligning opposing fibers. The inventor of the opposed-lens alignment concept was Melvyn A. Holzman.

Transfer-Molded Alignment Concept

Pairs of fiber end faces are highly polished to prevent Fresnel reflections, then insert-molded on the axis of the resulting ferrule utilizing a fabricated and very precise tool (see figure 4–6). The two mating ferrules (plugs) insert within a biconical socket. Within the socket are cushions that are index matched to accommodate the plugs. This concept was defined by

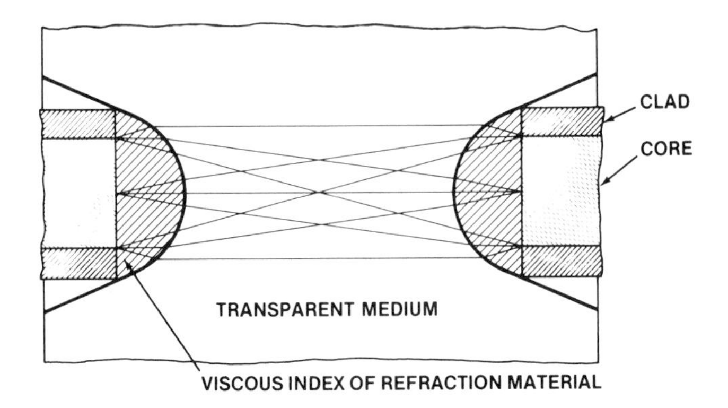

Figure 4–5 The Opposed-Lens Alignment Concept *(Courtesy ITT Cannon Electric)*

P. K. Runge and S. S. Cheng in the *Bell System Technical Journal,* July–August 1978.

Multirod Alignment Concept

In figure 4–7, six precision rods are located around the inside perimeter of a guide sleeve. Three extremely precise ferrule rods provide a mounting for the fiber. An end view of the assembled ferrule and guide sleeve shows the fiber inserted at the exact geometric center within the three-rod ferrule

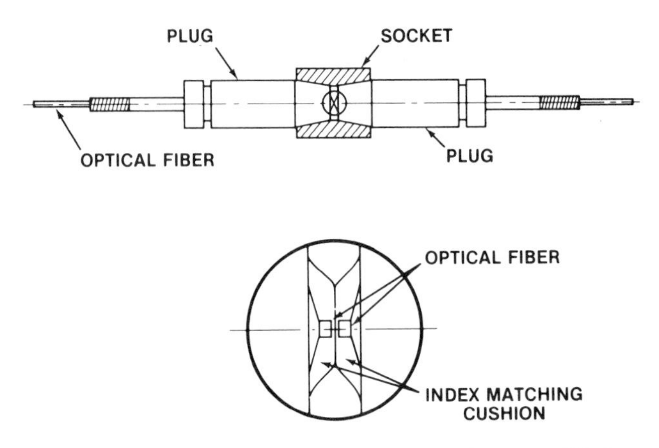

Figure 4–6 The Transfer-Molded Alignment Concept *(Courtesy ITT Cannon Electric)*

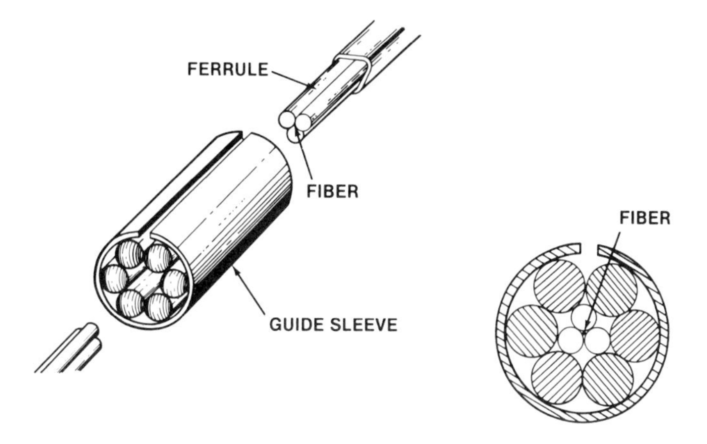

Figure 4–7 The Multirod Alignment Concept *(Courtesy ITT Cannon Electric)*

triangle. All rods are calculated for tangency using needle bearings or cam rollers. Tolerances using this method can be held to millionths of an inch. The inventor of the multirod alignment concept was Carl R. Sandahl.

Resilient Self-Centering Alignment Concept

Two fibers of slightly different diameters may be aligned using the resilient self-centering alignment concept. In figure 4–8, the end view illustrates the fiber installed in a V-groove that establishes three planes of force on the fiber. Force is transferred equally to three arcs on the perimeters of the

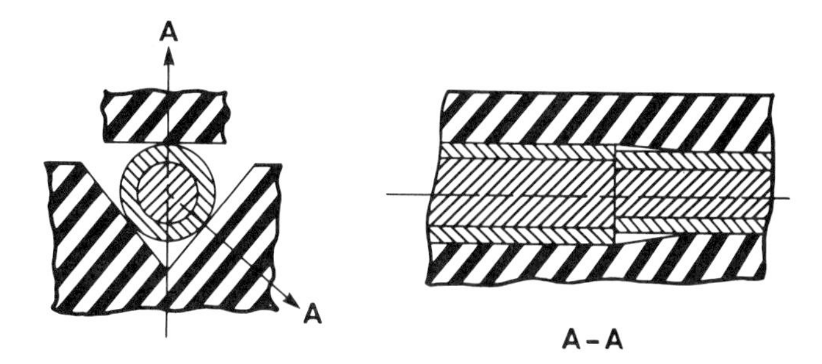

Figure 4–8 The Resilient Self-Centering Alignment Concept *(Courtesy ITT Cannon Electric)*

fibers. The material surrounding the fibers is resilient and will remember its original shape if the fibers are removed.

Double Eccentric Alignment

The double eccentric method of fiber alignment is one of the standard alignment procedures. The procedure deals with an eccentric motion similar to X–Y mechanisms. Adjustment is made in one direction or the other until fibers align. In figure 4–9 a pair of opposing ferrules is inserted into a guide cylinder and/or a male-female arrangement. The ferrules are utilized to adjust the fiber core to the axis of the outer cylinder surface. Adjustment is made under a microscope. When aligned, the ferrules are locked into place. The fiber ends are ground and polished. Epoxy is used to bond the fiber into the center sleeve.

A second method used with the double eccentric concept is to mate the fibers into a system, then adjust the eccentric until optimum power is realized at the receiver. The double eccentric concept was invented by Setsuo Sato.

Watch Jewel Concept

Readily available watch jewels are utilized to align fibers precisely, as in the arrangement shown in figure 4–10. The jewels are produced at reasonable prices, with 3-μm concentricity inside or outside diameter. The alignment process involves measuring the diameter of the fiber with accuracy, then choosing the jewel to fit that is just larger in size than the fiber. The fiber is then epoxied in place and ground and polished to tolerance. The

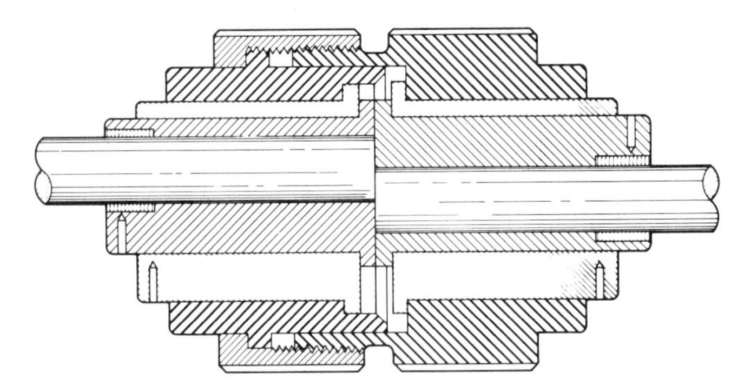

Figure 4–9 The Double Eccentric Alignment Concept *(Courtesy ITT Cannon Electric)*

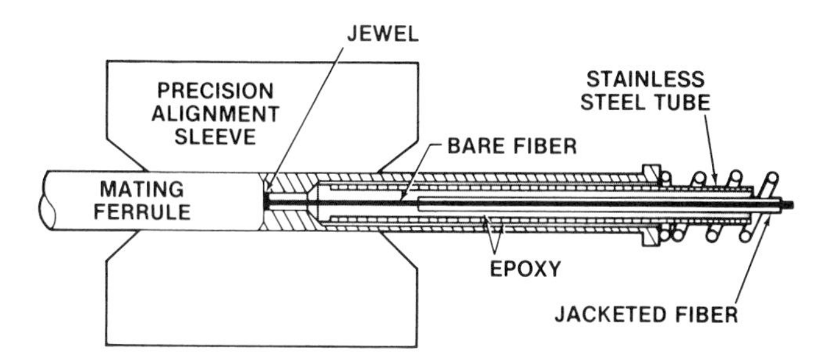

Figure 4–10 The Watch Jewel Alignment Concept
(Courtesy ITT Cannon Electric)

second fiber end is treated in the same manner. Finally, the two opposing jewel/ferrules are mated in a guide sleeve as pistons in a cylinder.

Three-Sphere Concept

Three spheres of exact diameter are placed in a pair of ferrule connectors (see figure 4–11). Each connector utilizes the three spheres in a plane of 120° increments with the spheres retained in a spherical race. In the center of the three spheres, a defined space for the fiber end is positioned. When the two pairs of ferrules are brought together, the two sets of spheres nest with respect to each other at 60° increments, precisely aligning the optical fibers. Alignment accuracy is influenced only by the variation in sphere diameter. Spheres are available with ten-millionths of an inch tolerance.

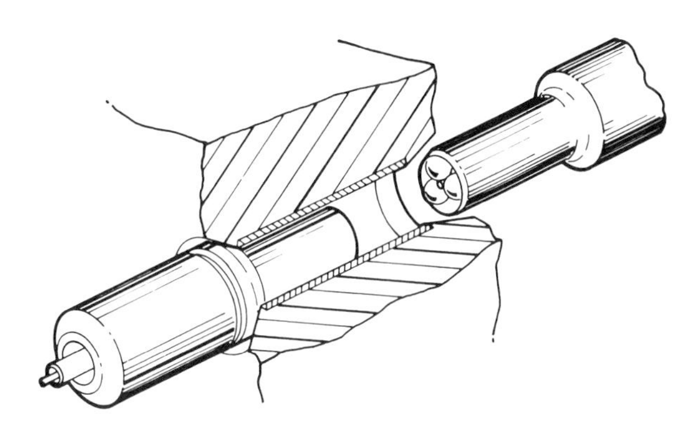

Figure 4–11 The Three-Sphere Alignment Concept *(Courtesy ITT Cannon Electric)*

5

Optical Fiber Manufacturing

GENERAL TECHNIQUES

The manufacturing techniques involved in the production of optic fibers are highly sophisticated processes. Most of these processes are secret or at the least proprietary. Considering the competitive nature of the product, this is understandable. There are three general techniques used in the fabrication process. The first to consider is to melt glass in one or two containers and pour the glass together to form the fiber. The second is: create a glass core or cladding, then melt a second glass to liquid or glass and deposit the second as a core or cladding on the original. The third is to begin with an exceptional rod of glass and coat it with a plastic cladding. The two processes that we shall discuss are the double-crucible and the chemical-vapor-deposition (CVD) processes.

Double-Crucible Method of Fiber Fabrication

In early processes, the rod-in-tube method was used (see figure 5–1). This method exudes cladding glass over core glass. The double-crucible furnace is built so that two crucibles holding cladding glass and core glass are vertically aligned. The crucibles are made from platinum or other heat-resistant material. Simple silica crucibles may be used, but they have a tendency to shatter when cooling to room temperature. Each crucible has a narrow aperture at its base. As the cladding glass and core glass are heated, they flow symmetrically through the crucible apertures so that a composite glass flow is produced. The composite glass may be pulled to form a perfect or nearly perfect fiber. The dimensions of the two crucibles

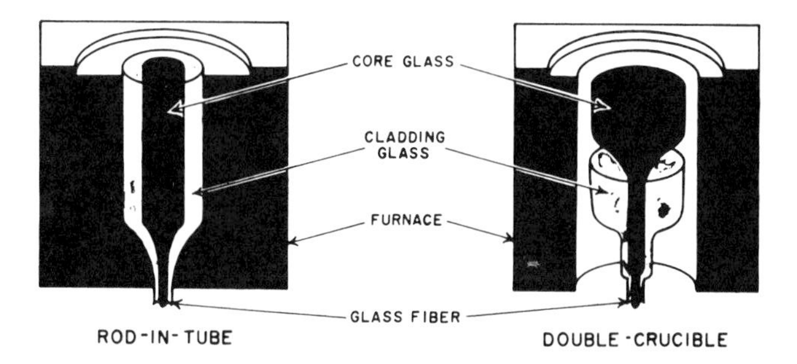

Figure 5–1 Double-Crucible Method of Fiber Fabrication *(Courtesy Corning Glass Works)*

are very exact, to ensure that core-to-cladding ratios are met. The viscosities and densities of the materials must be considered in flow restrictions, but usually are very similar in both cladding and core. The atmosphere around the crucibles is carefully controlled to prevent contamination and introduction of oxygen, which will cause bubbles in the glass and consequent high scatter losses.

Chemical-Vapor-Deposition (CVD) Process

The CVD method was first used by Corning Glass Works of Corning, New York, when they developed the first 20-dB/km fiber. The technique was called the *soot process*. Figure 5–2 is a line drawing of two doped-depos-

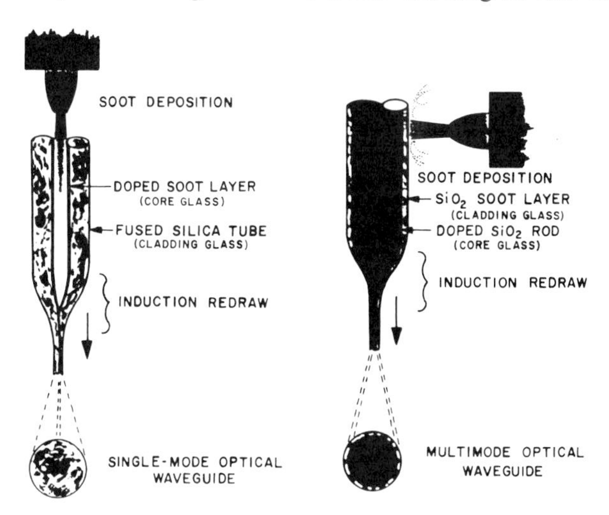

Figure 5–2 Inside and Outside Vapor-Phase-Oxidation Process *(Courtesy Corning Glass Works)*

ited-silica fabrication processes. On the left is an inside vapor-phase-oxidation process. This process is used to create single-mode optical waveguides. The outer layer of glass is a fused silica tube used as the cladding. Gaseous components react in a flame to produce a fine deposit of doped or undoped silica. The soot deposition is internal in this case. It is deposited as a doped layer to form the core. The redraw process involves heating the glass by induction and drawing it to form the single-mode optical waveguide. On the right in figure 5–2, the outside vapor-phase-oxidation (OVPO) process is illustrated. The inner layer provides the core glass. The outer soot layer (that was deposited), establishes the cladding. The redraw process involves heating the glass by induction and drawing it to form the multimode optical waveguide.

The doped-deposited-silica (DDS) process shown in the figure 5–3 allows the introduction of carefully controlled amounts of dopants to the silica tube. The dopants may be increased or decreased to raise or lower the refractive indexes of the host glass, which carefully defines the waveguide (fiber). This makes the DDS process uniquely suited for graded-index

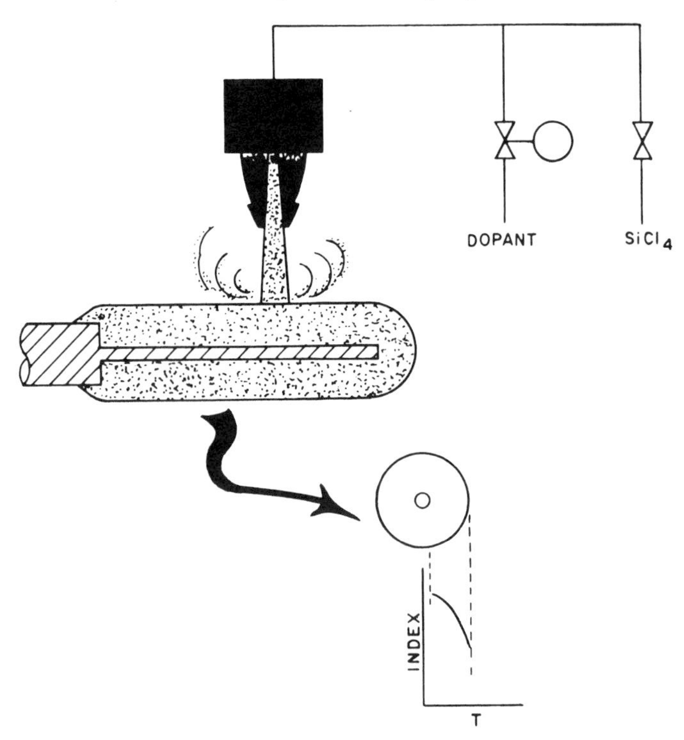

Figure 5–3 Dope-Deposited-Silica Process *(Courtesy Corning Glass Works)*

waveguide fabrication. Graded-index waveguides require a continuously varying, precisely controlled profile of the refractive index within the waveguide core.

A second class of methods leading from soot to chemical vapor deposition produced a glassy layer rather than sootlike layer on the surface of a silica tube or the inside of the tube.

The CVD process is by far the most widely utilized process in developing and manufacturing of waveguides. The process begins with an extremely high-quality tube of quartz, such as that shown in the figure 5–4.

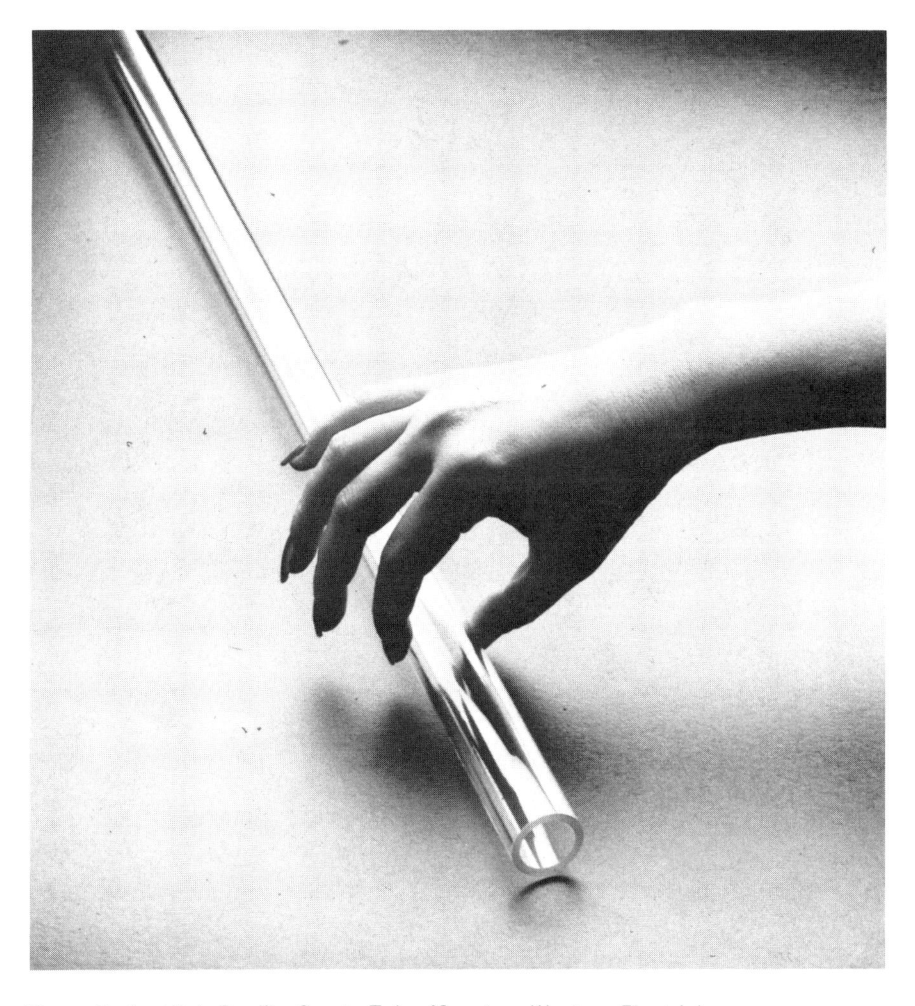

Figure 5–4 High-Quality Quartz Tube *(Courtesy Western Electric)*

The glass tube's length is near a meter and its diameter is about 20 mm, depending on the fiber being processed and the manufacturing technique. The tube is placed in a special lathe (see figure 5–5) to create a preform which has the properties of the fiber. Heat is applied. A constant gas flow through the tube deposits boron-doped silica on the inner wall of the tube. The boron-doped silica has a lower refractive index than pure silica. This process provides a core material for the fiber. The borosilicate soot is uniformly deposited by the maintenance of a constant level of gas flow and a regulated flame along the tube. By repeating the process with varying amounts of dopant, graded- or step-index core regions may be built up. Germanium and phosphorus dopants are used to increase refractive indexes. Boron is used to lower refractive indexes.

Figure 5–6 is a quartz glass tube that has been through the deposition

Figure 5–5 Preform Lathe *(Courtesy Western Electric)*

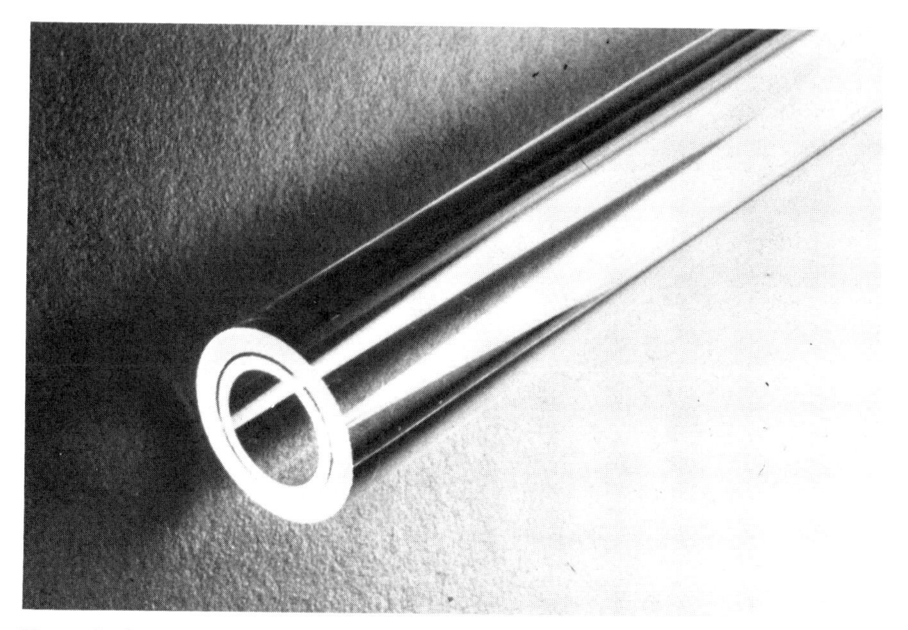

Figure 5–6 Quartz Tube After Deposition Process *(Courtesy Western Electric)*

process. After a suitable number of layers of borosilicate has been deposited, the tube is collapsed to form the fiber preform (see figure 5–7). The preform is then placed into a specially designed lightguide drawing machine where the quartz preform is heated in an induction furnace or by CO_2 laser. In the machine, the preform is pulled into several kilometers of hair-thin lightguides. The lightguide drawing machine is a precision instrument. Figure 5–8 is a photograph of one of the drawing machines developed by Bell Laboratories and Western Electric.

It must be noted that the fabrication processes described here are simplified, because the scope of this book is introductory. The mechanics of fiber fabrication are critical and complex. Glass melting, flow rate, diffusion methods, geometry of containers, cooling, crucible loading and feeding procedures, atmosphere control, pulling/drawing, and winding are just a few of the exacting details involved in the fabrication process.

OPTICAL FIBER MATERIALS

Most fibers are made from fused mixtures of oxides, sulfides, or selenides. The optically transparent mixtures used in fibers are called *oxide glasses.*

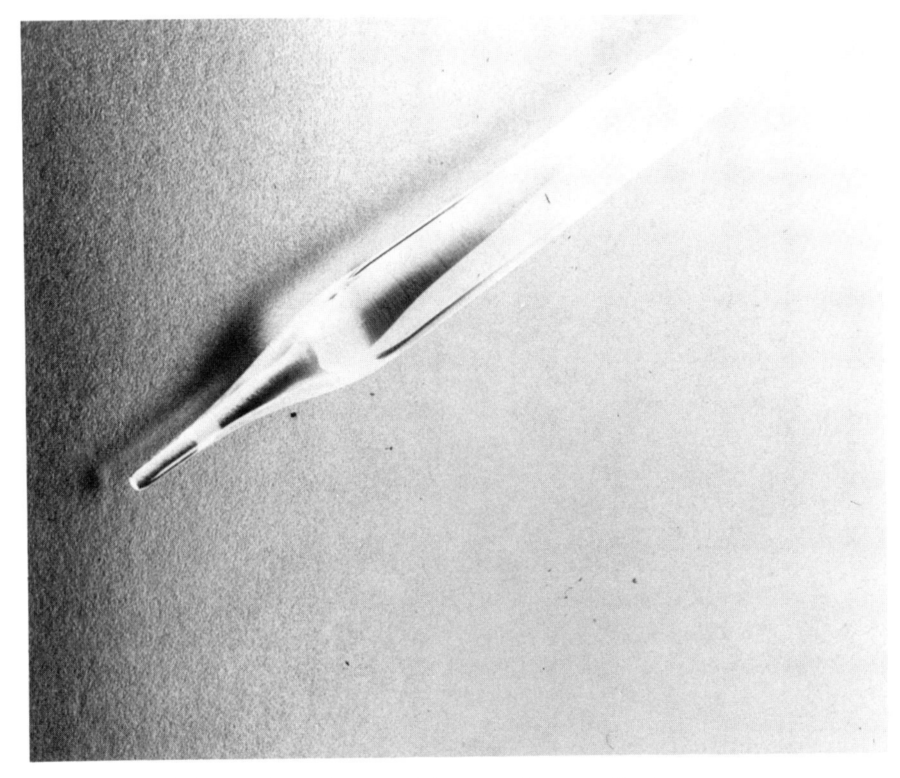

Figure 5–7 Collapsed Fiber Preform *(Courtesy Western Electric)*

The oxide glasses are common silica (SiO_2); lead silicate, such as crystal; and sodium calcium silicate, such as plate glass. Materials involved in the process are sodium carbonate, calcium carbonate, boric oxide silica (sand), and lead oxide. The chemical mixtures are extremely complex and must satisfy all the requirements of attenuation, strength, and so on, that were just discussed.

The processes used will produce single fibers in one of four different forms:

1. Plastic core with glass cladding
2. Glass core with glass cladding
3. Plastic core with plastic cladding
4. Glass core with plastic cladding

The selection of the correct material does not always have to do with the desired refractive indexes. Softening temperatures of the materials should be similar. Both the core and the cladding are heated in the same furnance. If they do not soften at similar temperatures, drawing conditions may not be right. Thermal expansion of the core should be greater than that of the cladding. This helps to increase the strength of the fiber. Different temperatures may cause adverse reaction at the interface.

Figure 5–8 Light-Guide (Fiber) Drawing Machine *(Courtesy Western Electric)*

6

Typical Optical Fibers and Cables and Their Installation

Optical fibers are transparent, dielectric cylinders surrounded by a second dielectric cylinder. There are a great number of types on the market. Each is manufactured for some specific need such as dispersion, attenuation, bandwidth, and tensile strength. In this chapter we look at fibers and cables together with some of their properties and installation.

Cables are constructed usually for the ability to protect the fibers within from ambient dangers such as water, humidity, heat, and cold. Fibers and cables are first of all obviously selected for specific parameters of transmission. This goes without saying.

FIBER TYPES

Fibers are constructed from plastic or glass or a combination of the two; for example, glass core/glass cladding, plastic core/plastic cladding, or glass core/plastic cladding. The choice of a fiber type has to do with the quality, the parameters required, and economics. Can the system live with dispersion or must we use a better fiber at a higher price? In real life there are trade-offs in design. You cannot always receive lower cost without lower performance. A lower attenuation per kilometer may also mean a very high price tag.

In the next several paragraphs we shall take a look at several fiber types. As you are comparing fiber types, consider what makes each more or less desirable than others.

Fiber Transmission Types

1. The single-mode step index designated is designed for use in very high bandwidth (500 MHz or more) single-fiber data systems or in other applications that require single-mode propagation of light.
2. The glass step-index fiber is designed for use in medium-to-high-bandwidth single-fiber data transmission systems.
3. The glass graded-index fiber is designed for use in high-bandwidth single-fiber data transmission systems.
4. The wideband glass-graded index is designed for use in applications demanding the large channel capacity.
5. Large-core plastic-clad silica fiber is designed for use in moderate-distance, medium-bandwidth single-fiber data transmission systems.
6. Plastic-clad silica fiber is designed for use in economical, medium-bandwidth, medium-distance, single-fiber data transmission systems.

Fiber Specifications Fiber specifications were previously discussed in detail, however, we list the 10 major specifications here for review:

1. Attenuation
2. Intermodal Dispersion (−3dB)
3. Core Index of Refraction
4. Fiber Core Diameter
5. Fiber Core Outer Diameter
6. Jacket Outer Diameter
7. Tensile Strength
8. Minimum Bend Radius
9. Numerical Aperture
10. Bandwidth Parameters

Single-Fiber Crossections In Figure 6–1A a crossection of a single fiber is shown. This fiber has a large core which is particularly suitable for medium bandwidth transmission. Large cores allow more efficient coupling for high power levels at the receiver. The fiber has a large core, a plastic cladding, and protective plastic jacket. This fiber could be placed in a cable by itself or in a multiple-fiber cable.

Figure 6–1B has a doped silica core and a borosilicate cladding. This profile is best suited for very high bandwidth and with single-fiber single-mode transmission. The fiber has an inner and an outer jacket to ensure mechanical strength and environmental protection.

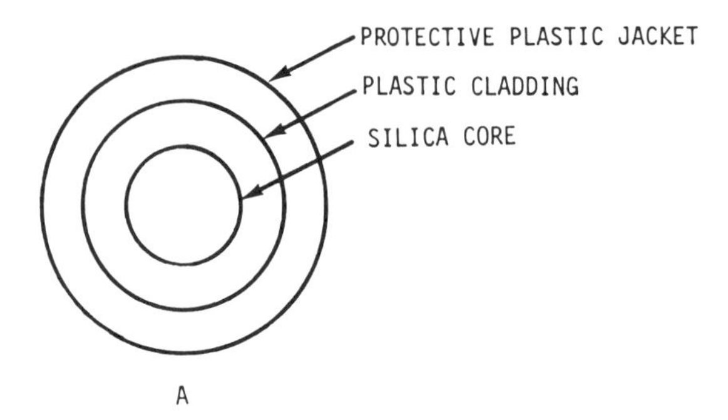

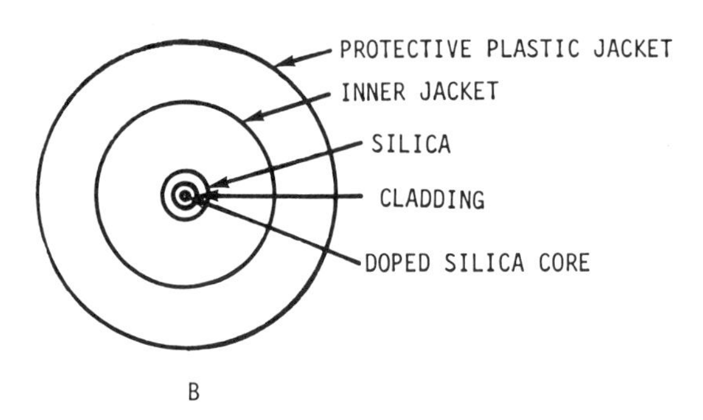

Figure 6–1 Fiber Crossections

CABLE TYPES AND SPECIFICATIONS

Cables of single or multiple fiber are manufactured usually for protection of the fiber. Fibers are encased within the cable using protective jackets (usually polyurethane). Strength members surround the fibers. Cables are constructed to feature strength, light weight, high flexibility, resistance to kinks, and resistance to crushing. Some tests imposed on cables in the development stage are impact, bend, twist, fatigue under load, high and low temperature, and even storage losses.

As in the choice of fibers, certain trade-offs are often necessary. However, the reasons for cable development are centered around strength and environmental conditions. These should not be traded-off in the search for economics. The next several paragraphs contain information on cable types. The reader should compare each type and determine what the benefits are for using a particular cable.

Cable Specifications As with fibers, cables are also specified. Cables have first of all, the fiber optical parameters which must be met and then the mechanical and dimensional specifications. Some of these are as follows:

1. Tensile strength: usually in Newtons
2. Bend radius: in centimeters
3. Crush resistance: in Newtons/centimeters
4. Impact: Hammer or other device and number of strikes
5. Flexing: number of times and angle of flexing
6. Numbers of fibers: 1,2,3, etc.
7. Weight: in kilograms/kilometers
8. Length: in meters
9. Shape: when customized to fit a specific geometry

Special Parameters The fiber must be made to transmission specifications. However, there is some limit which must be maintained in order to meet these specifications. The cable designer with enough time and money can be creative enough to provide for excesses in temperature or humidity, strength, non-symmetric and unusual shapes, special flammability requirements, and toxicity concerns.

However, there are limits in cable design also. For instance, it is not possible to have extreme light weight and extreme strength. Nor is it likely that the designer can produce a miniature package with a multiple of fibers. Most of this is common sense.

Cable Problem Areas Although we have discussed fiber parameters at length throughout the book, there are some areas that relate especially to cable installation and therefore design. Three major problem areas that must be carefully considered are as follows:

1. Microbends
2. Attenuation with Temperature
3. Elongation of the Fiber

Microbends are miniature bends in the fiber axis due to physical contact with the structural elements in the cable. These are cared for by placing a loose-fitting buffer jacket around the fiber to isolate it from other members of the cable. The fiber may then move rather freely in the space without strength members effecting it when they are under any stress.

The second problem area is related to microbends, in that attenuation changes with temperature. In optical fibers, this is due to mismatch of expansion coefficients of glass and plastic components which cause microbending. Temperature microbends are cured by loose-fit decoupling of strength members from the fiber itself.

The third problem area is elongation. Any stress placed on the cable may effect the fiber. The shorter the cable, the less problem we have in this area. The longer the cable the greater chance of stretching, breaking, causing surface flaws, and other disturbances which affect the basic fiber parameters.

Cable Crossections Figure 6–2A shows the simplest of the cable types. Each cable manufacturer has some version of the single fiber. They are all similar to this type. The fiber is placed in a buffer tube or jacket. The fiber may be tightly packed to isolate from external damage or influence from its environment. This method does, however, have some optical loss.

The fiber in this buffer tube (Figure 6–2B) may also be loosely buffered. That is, the buffer tube or jacket may be somewhat larger than the fiber outside diameter. This allows the cable to lengthen under load without stress to the fiber.

In Figure 6–3A, a simple cable is constructed in three layers. The buffered tube surrounds the fiber in a loose packed arrangement. External to the tube are up to ten strands of fiberglass. The fiberglass strands are used to carry the load and to allow parallel stiffness to the cable and fiber. The fiberglass is covered by a polyurethane jacket which provides cushioning.

In Figure 6–3B a Kevlar fibered jacket is placed around the polyurethane for tensile strength.

Still another version of the cable is that shown in Figure 6–3C. A further layer of polyurethane or PVC is used as a sheath.

Figure 6–2 A Single-Fiber Buffered Structure

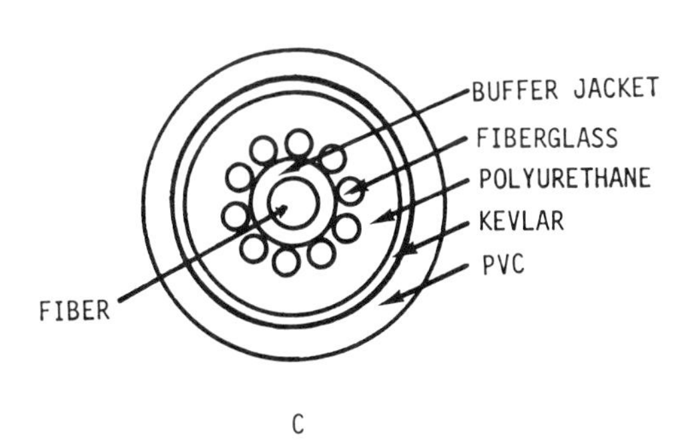

Figure 6–3 Single-Cable Construction

Plow-In Cables There is very little deterioration of optical qualities when bare fibers are exposed to moisture without extremes. However, when water penetrates the jacket at temperature below 0°C, it leads to ice forming inside the buffer which, in turn, creates attenuation due to the expansion of the ice and microbending. To avoid these effects the buffer jackets are filled with a substance called polyurethane jelly. The jelly is soft and pliable to allow fiber movement.

In Figure 6–4, a plow-in cable with 6 fibers and a filled buffer is shown. This cable was manufactured by Siecor for the United Cable Company of Pennsylvania. The cable is a direct burial 6-fiber optical type. It was used in an installation between Carlisle and Mt. Holly Springs, Pennsylvania. Each pair of fibers is capable of handling up to 672 voice channels operating at 44.7 Mbit/second, which corresponds to the third level (T3) of the digital telephone hierarchy. The distance was 12.3 km in the interoffice trunk system in Carlisle. The reader can see the structure of the cable in Figure 6–4.

In Figure 6–5, the 6-fiber cable is compared with a conventional telephone cable which does the same task.

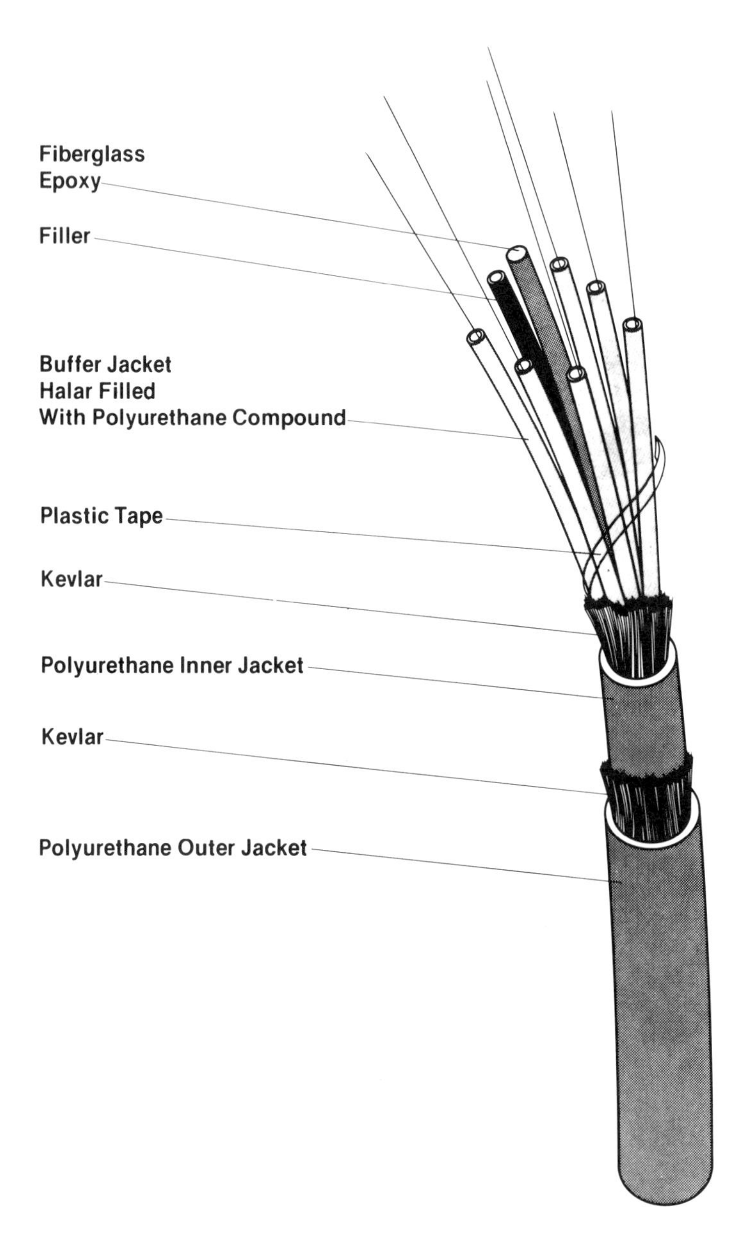

Figure 6–4 Plow-In Cable Used in Carlisle, Pa., with a United Telephone Company of Pennsylvania Interoffice Trunkline *(Courtesy, Siecor Optical Cables, Inc.)*

Figure 6–5 Comparison of 6-Fiber Cable with Conventional Telephone Cable *(Courtesy, Siecor Optical Cables, Inc.)*

Duct Installation Cables (See Figure 6–6) A 10-fiber buffer-filled optical cable was selected by the General Telephone Company of Indiana for a 5-kilometer interoffice trunk system in Fort Wayne. The cable is 5/16th of an inch in diameter. Distance of the trunkline is 2.8 mile (4.5 kilometer). The link services as a inter-office trunk and will handle other communication services.

The cable contains 10 optical fibers. Each of the 10 optical fibers is surrounded by a viscous polyurethane compound that insures that the mechanical and optical properties of the fibers remain constant in a hostile environment. Each fiber pair can provide 672 voice channels of telephone communications. Note that the cable has a fiber glass rod running through its length. This was substituted for the usual steel wire.

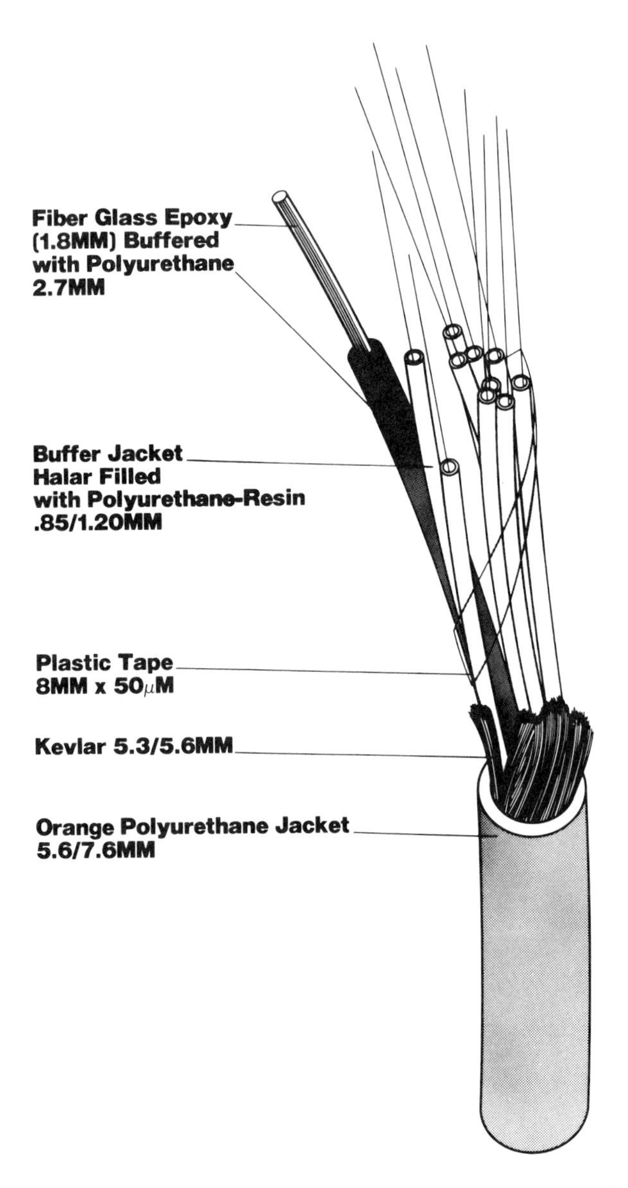

Figure 6–6 A 10-Fiber Cable (Filled Buffer) for use with a Duct Installation in General Telephone of Indiana's Fort Wayne Trunkline *(Courtesy, Siecor Optical Cables, Inc.)*

Aerial Cables When applications of cables must stretch over a length of space between telephone poles and the like, one must consider heavy loadings of radial ice. In the northern hemisphere, operating conditions may range from −40°C to +50°C. Under these conditions, suitable support for this cable demands that steel spring wire be used as a strength member. The number and diameter of the enclosed steel wire is dependent of course on the load. The load is usually determined by the distance between connections and the anticipated sag.

This particular cable in Figure 6–7 has 4 graded-index optical fibers and 2 copper wires.

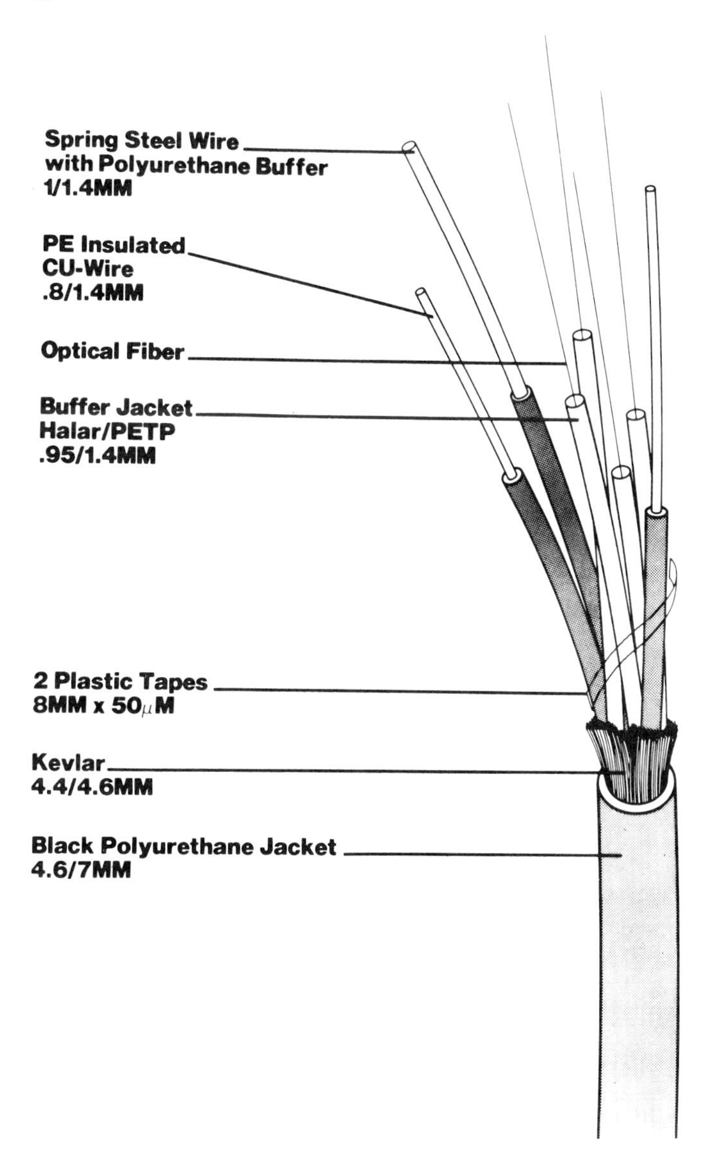

Figure 6–7 Lashed Aerial Optical Cable *(Courtesy Siecor Optical Cables, Inc.)*

Each fiber is enclosed in a loose buffer tube filled with a viscous polyurethane compound to protect it from moisture and freezing. The cable construction begins with a steel wire central member around which the buffered fibers and copper wires are helically stranded. The spaces between the fibers, copper wires, and steel strand are also flooded with the polyurethane compound. This cable core is reinfored with layers of aramid yarn, and enclosed in a polyurethane outer jacket.
The cable specifications are:

Maximum Attenuation @ 820 μm	6.5 dB/km
Bandwidth @ 1 km (−3 dB optical)	500 MHz
Cable Weight	64 kg/km
Operating Temperature Range	−35°C/+50°C
Radial Ice Loading	13 mm
Tensile Strength	1000 N (225 lbs.)
Reel Length	3 km

Ribbon Cable (See Figure 6–8)

The ribbon cable is designed for parallel or multi-channel data transmission applications. This configuration can be used in trays and conduits for intrabuilding systems. The attenuation of the cable is <10dB/km and has a bandwidth >200MHz•km. The ribbon cable is usually customized with any number of fibers/channels to meet various user requirements.

The flat nature of the ribbon cable allows it to be fitted and attached to flat surfaces where they will not be seen. This is a desirable feature for intra-building use.

External-Strength-Member Heavy-Duty Optical Fiber Cable

The external-strength-member heavy-duty optical fiber cable is designed for use in single-fiber-per-channel transmission systems which require high cable strength and crush resistance. In addition, its high flexibility and kink resistance makes it ideally suited for use in conduit, cable trays, and a variety of intravehicle applications.

FIBER ENVIRONMENTAL INFORMATION

Operating temperature range for standard installed cable is −20 to +50°C. If cables are located where there is a possibility of immersion in water at temperatures below 0°C, special cables with a polyurethane filling

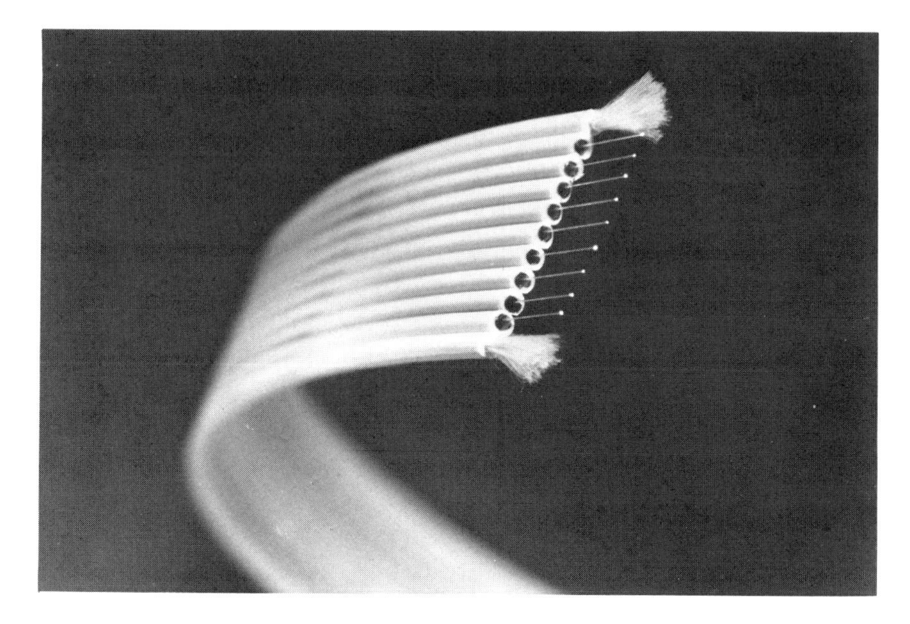

Figure 6–8 Ribbon Cable *(Courtesy Siecor Optical Cables, Inc.)*

compound in the buffer tubes can be used to prevent water seepage. This is because of the possibility that damage to the cable jacket could result in water leaking into the buffer tube. If this water freezes, expansion of the ice could cause microbending of the optical fiber, which could increase optical attenuation, and possibly damage the fiber. The polyurethane filling serves the same purpose as the pressurization usually employed to protect wire cables from water, and is much simpler and requires no maintenance.

Cables for extreme high and low temperatures are also available. For use in environments where the fiber-optic cables may be exposed to fire, cables that meet the IEEE-383 flame-retardance specification can be supplied on special order.

FIBER-OPTIC CHEMICAL RESISTANCE PROPERTIES

Fiber-optic cables are superior to copper wire cables for use in the presence of corrosive chemicals for two reasons: damage to the fiber-optic cable jacket by any chemical other than hydrofluoric acid (the only chemical which attacks glass at ordinary temperatures) will not interrupt data transmission, and no sparks can result from any type of damage to the cable. However, damage to the Kevlar and polyurethane layers will cause mechanical weakening of the cable, which could result in damage to the optical fiber from mechanical stresses. A qualitative rating of resistance to common types of chemicals for both polyvinyl chloride (PVC) and polyurethane is given in Table 6–1. Cables with polyurethane outer jackets are available on special order for applications where characteristics of this material are more desirable than those of PVC. More detailed information can be provided if required.

TABLE 6–1 Resistance Ratings*

	PVC	*Polyurethane*
Oxidation resistance	E	E
Heat resistance	G-E	G
Oil resistance	E	E
Low-temperature flexibility	P-G	G
Weather, sun resistance	G-E	F-G
Ozone resistance	E	E
Abrasion resistance	F-G	O
Electrical properties	F-G	P-F
Flame resistance	E	P
Nuclear radiation resistance	P-F	G
Water resistance	E	P
Acid resistance	G-E	F
Alkali resistance	G-E	F
Gasoline, kerosene, etc. (aliphatic hydrocarbons) resistance	G-E	F
Benzol, toluol, etc. (aromatic hydrocarbons) resistance	P-F	P
Degreaser solvents (halogenated hydrocarbons) resistance	P-F	P
Alcohol resistance	G-E	P

*P, poor; F, fair; G, good; E, excellent; O, outstanding.

NUCLEAR RADIATION

Fiber-optic cable is better than wire for use in nuclear power plants and other high-radiation-risk areas for the following reasons. The cable is nonconductive and, unlike copper wire, does not build up static electric charge in the presence of radiation. Fiber-optic cable will not short out if its jacket is melted by the high heat of a runaway nuclear reaction, ensuring continued communication and control in this type of emergency. Since optical fibers are much smaller than most wires, it is easier to make fiber-optic cables self-sealing to keep the cable from becoming a hose that would allow radioactive gases to escape from a sealed area.

If fiber-optic cable must be used in an area where it is subjected to continuous high-intensity radiation, another effect must be considered. Nuclear radiation causes increases in light absorption by impurities in the optical fiber, which in extreme cases can reduce the received optical signal power to an unusable level. Maximum allowable loss in the fiber-optic cable is 10 dB. Since radiation increases loss per unit length, maximum allowable radiation dose is a function of cable length. (Or, alternatively, maximum cable length is a function of radiation dosage that must be tolerated.)

These losses are highly dependent on radiation intensity, as well as total dosage. Research results presently available indicate that a very high intensity burst (3700 rad in 3 ns) produces an optical attenuation peak of thousands of dB/km, which decays in less than 10 s to 50 dB/km (usable length reduced to 200 m). For low-level, long-term irradiation, cable attenuation is doubled (standard 1-km usable length reduced to 500 m) by a cumulative radiation dose of approximately 200 rad, and usable length is reduced to 200 m by a dose of 1200 rad. These values are dependent on temperature and radiation intensity, and possibly other factors. Research is presently being conducted to determine more accurately the effects of radiation on optical fibers and to improve the radiation resistance of the fibers.

MECHANICAL PROPERTIES

All cables, both wire and fiber optic, have a maximum tensile load and minimum bend radius. For wire cable, minimum bend radius is determined by the cable size and type of construction. High-frequency coaxial cables, such as RG/8 or RG/11 (attenuation at 200 MHz = 72 dB for 1 km) have an outside diameter of 0.4 to 0.5 in. (10 to 12.5 mm) and a minimum bend radius of approximately 4 to 5 in. (100 to 125 mm). Coaxial cables in the same physical size range as standard fiber-optic cables such as RG/58 (0.195 in. or 4.95 mm outside diameter) typically have recommended maximum tensile loads in the range 60 to 100 lb (267 to 445 newtons).

The minimum bend radius and maximum tensile load parameters of a fiber-optic cable interact, and both have an effect on the optical characteristics of the cable. An increasing tensile load will cause: first, a reversible increase in attenuation; second, an irreversible increase in attenuation; and finally, cracking of the optical fiber. With a smaller bend radius, these effects occur at lower tensile loads. Under no circumstances should the cable be bent in a curve with a radius of less than 1.2 in. (30 mm). During installation, to guarantee no permanent effects on cable optical properties, maximum tensile force should be limited to 90 lb (400 N). Minimum bend radius with this tensile force applied is 5.9 in. (150 mm). In operation, to ensure against degradation of optical transmission properties of the fiber, maximum tensile force is 11.25 lb (50 N). Minimum bend radius with no tensile force applied is 30 mm. If any tensile force is applied, the minimum bend radius is 2.0 in. (50 mm) for forces up to 1.1 lb (5 N), 3.0 in. (75 mm) for forces up to 2.2 lb (10 N), and 4.0 in. (100 mm) for forces between 2.2 and 11.2 lb.

The other mechanical stresses that may be of concern in some installations are crushing and flexing. Maximum crushing force (evenly distributed over parallel surfaces) to which the cable can be subjected without any mechanical damage is 57 lb/in. (100 N/cm) of cable length to which it is applied. Forces of 57 to 114 lb/in. (100 to 200 N/cm) will cause deformation of the buffer tube, but no effect on optical properties. Maximum crushing force that will not cause any irreversible degradation in optical properties is 228 lb/in. (400 N/cm). The standard flexing test of the cable consists of wrapping it in a 180° arc around a 50-mm radius, alternating in opposite directions, 6000 times with a tensile force of 0.225 lb (1 N) applied to the cable. Some cable samples have been subjected to this test up to 30,000 bends without any damage. However, it is not recommended that the cable be subjected to this type of stress continuously. If it is necessary to flex the cable continuously, or on a regular long-term basis, it is recommended that a larger radius be used.

All the mechanical stress limits discussed in the preceding paragraph are to some degree interdependent on each other, and on temperature. Since it would be prohibitively expensive to test all possible combinations of these limits, any user considering operating near the extremes of two or more mechanical characteristics, particularly at extremely high or low temperatures, should arrange for special tests of the cable, or use one of the high-strength cable types designed for aerial or burial use.

CABLE ROUTING

Fiber-optic cables may be placed in trays or ducts, pulled through conduit, suspended in the air, or buried underground. For direct burial, special cables are available which are able to withstand the tensile and crushing forces imposed by plowing in and by earth displacements after installation, and are protected against possible attack by burrowing or gnawing animals. Standard cable can be used under ground under the following conditions. The cable must be located below the frost line and in an area where there is no possibility of water seepage and no possibility of attack by animals. Cable should be enclosed in a polyethylene or PVC pipe with inside diameter at least four times as large as the outside diameter of the cable to protect against the effects of earth movements, and an excess length of cable should be included inside the pipe to prevent the possibility of tensile loads being placed on the cable. Users considering buried cable installations should consult the factory for more detailed information about this type of cable and installation techniques.

As with copper wire, fiber-optic cables used in aerial installations also require special cables which are designed to withstand the stresses of ice and wind loading and temperature extremes. In mild climates it may be possible to use a standard cable lashed to a galvanized steel messenger wire to provide continuous support and protection against excessive tensile loading.

TRAY AND DUCT INSTALLATIONS (See Figure 6–9)

The easiest place to install fiber-optic cables is in trays and ducts. Since standard fiber-optic cables are electrically nonconductive, they can be placed in the same ducts with high-voltage cables without the special insulation which would be required by copper wire. The only ducts where fiber-optic cables should not be located are air-conditioning and ventilating ducts, for the same reason that PVC insulated wires should not be placed in these locations; a fire inside these ducts could cause the PVC outer jacket of the fiber-optic cable to burn and produce toxic gases.

The first mechanical property of the cable that must be considered in planning an installation is outside diameter, both of the cable itself and the connectors. The connector outside diameter is 0.38 in. (9.7 mm). The outside diameter of simplex cable is 0.189 in. (4.8 mm). The duplex cable has an oval cross section, 0.193×0.335 in. (4.9×8.5 mm). (*Note:* These dimensions apply only to standard one- and two-fiber cables and not to multifiber or other special types.) If the cable must be pulled through a conduit or duct, the minimum cross-section area required is 0.79×0.43 in. (20×11 mm) for duplex cable. For simplex cable, minimum cross-sectional area is determined by the pulling grip used. (Refer to the section "Pulling Fiber-Optic Cables.")

The primary consideration in selecting a route for fiber-optic cable

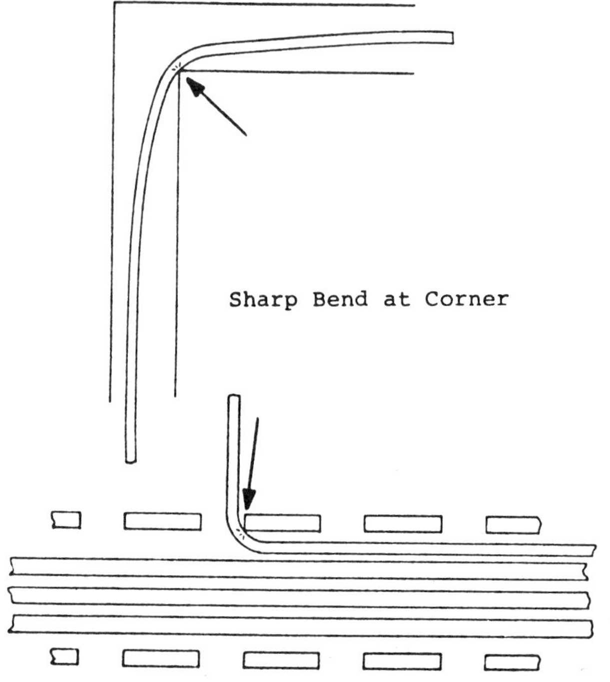

Figure 6–9 Cable Tray with Slotted Sides *(Courtesy Canoga Data Systems)*

through trays and ducts is to avoid potential cutting edges and sharp bends. Areas where particular caution must be taken are corners and exit slots in sides of trays.

If a fiber-optic cable is in the same tray or duct with very large, heavy electrical cables, care must be taken to avoid placing excessive crushing forces on the fiber-optic cable, particularly where the heavy cables cross over the fiber-optic cable. (See Figure 6–10.) In general, cables in trays and ducts are not subjected to tensile forces; however, it must be kept in mind that in long vertical runs, the weight of the cable itself will create a tensile load of approximately 0.016 lb/ft (0.24 N/m) for simplex cable, and 0.027 lb/ft (0.44 N/m) for duplex cable. This tensile loading must be considered to determine the minimum bend radius at the top of the vertical run. Long vertical runs should be clamped at intermediate points (preferably every 1 to 2 m) to prevent excessive tensile loading of the cable. The absolute maximum distance between clamping points is 330 ft (100 m) for duplex cable, and 690 ft (210 m) for simplex cable. Clamping force should be no more than is necessary to prevent the possibility of slippage, and is best determined experimentally since it is highly dependent on the type of clamping

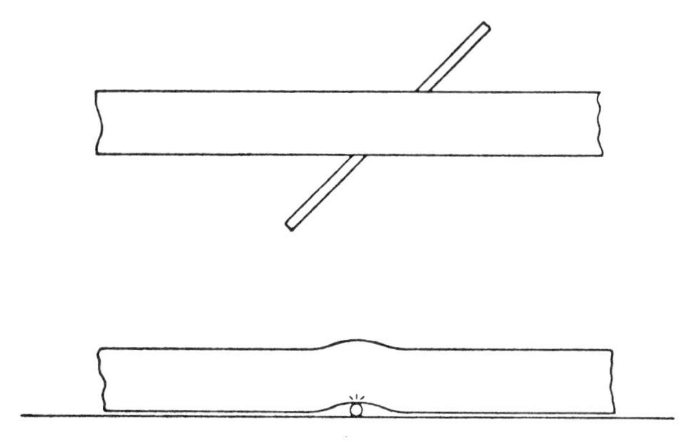

Figure 6–10 Fiber-Optic Cable Crossing Under Large Electrical Cable *(Courtesy, Canoga Data Systems)*

material used, and the presence of surface contaminants, both on the clamp and the jacket of the optical cable. As discussed in the "Mechanical Properties" section, the clamping force must not exceed 57 lb/in. (100 N/cm) and must be applied uniformly across the full width of the cable. The clamping force should be applied over as long a length of the cable as is easily practicable, and the clamping surfaces should be made of a soft material, such as rubber or plastic.

CONDUIT INSTALLATIONS

The first factor that must be considered in determining the suitability of a conduit for a fiber-optic cable installation is the availability of sufficient clearance between walls of the conduit and/or any other cables that may already be present to allow the fiber-optic cable to be pulled through without excessive friction or binding, since the maximum pulling force that can be used is 90 lb (400 N). (See cable dimensions in an earlier paragraph.) Since minimum bend radius increases with increasing pulling force, bends in the conduit itself and any fittings through which the cable must be pulled should not require the cable to make a bend with a radius of less than 5.9 in. (150 mm). Fittings, in particular, should be checked carefully to ensure that they will not cause the cable to make sharp bends or be pressed against corners. If the conduit must make a right-angle turn, use a fitting such as shown in Figure 6–11 to allow the cable to be pulled in a straight line and to avoid sharp bends in the cable.

Pull boxes should be used on straight runs at intervals of 250 to 300 ft to reduce the length of cable that must be pulled at any one time,

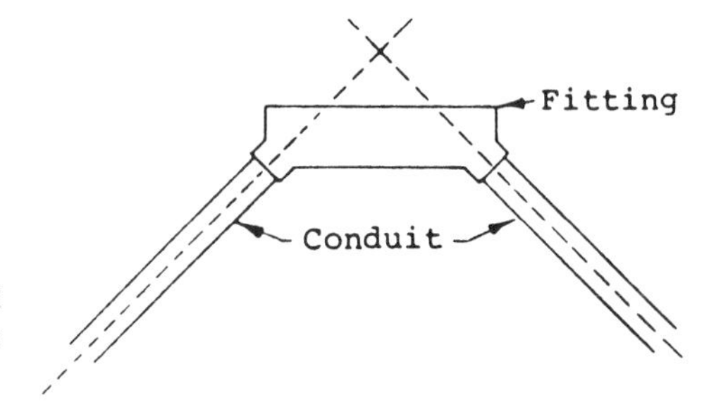

Figure 6–11 Fitting to Allow Straight Pulls at Corners *(Courtesy Canoga Data Systems)*

thus reducing the pulling force. Also, pull boxes should be located in any area where the conduit makes several bends that total more than 180°. To guarantee the cable will not be bent too tightly while pulling the slack into the pull box, the pull box must have an opening with a length equal to at least four times the minimum bend radius (4.75 in. or 120 mm). This is illustrated in Figure 6–12, which shows the shape of the cable loop as the last of the slack is pulled into the box. The tensile loading effect of vertical runs discussed in the section on tray and duct installations is also applicable to conduit installations. Since it is more difficult to properly clamp fiber-optic cable in a conduit than in a duct or tray, long vertical runs should be avoided, if possible. If a clamp is required, the best type is a fitting which grips the cable in a rubber ring. Also, the tensile load caused by the weight of the cable must be considered along with pulling force to determine the maximum total tensile load being applied to the cable.

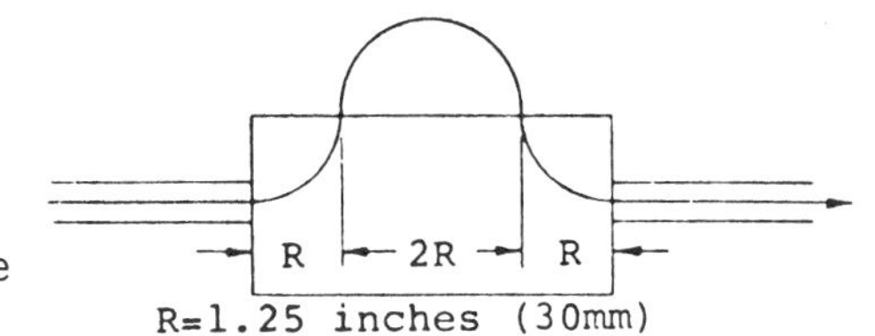

Figure 6–12 Minimum-Pull Box Size *(Courtesy Canoga Data Systems)*

PULLING FIBER-OPTIC CABLES

Fiber-optic cables are pulled in using many of the same tools and techniques that are used in pulling wire cable. The departures from standard methods are due to the fact that the connectors are usually preinstalled on the cable, the smaller pulling forces that are allowed, and the minimum bend radius requirements.

The pull tape must be attached to the optical cable in such a way that the pulling forces are applied to the strength members of the cable (primarily the outer Kevlar layer) and the connectors are protected from damage. The recommended method of attaching a pulling tape to a simplex cable is the "Chinese finger trap" type of cable grip. (See Figure 6–13.) The connector should be wrapped in a thin layer of foam rubber and inserted in a stiff plastic sleeve for protection. Since the smallest cable grip available (Kellems 033-02-044) is designed for 0.25-in. (6.4-mm) -diameter cable, and the outside diameter of the simplex cable is only 0.188 in. (4.8 mm), the cable grip should be stretched tightly and then wrapped tightly with electrical tape to provide a firm grip on the cable.

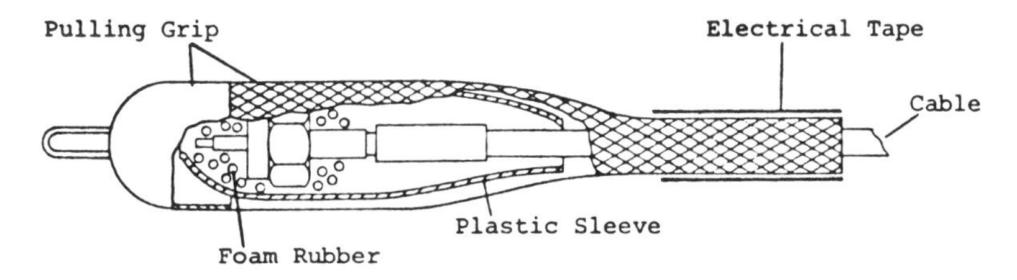

Figure 6–13 Attaching Pulling Grip to Simplex Cable *(Courtesy Canoga Data Systems)*

The duplex cable is supplied with the Kevlar strength members extending beyond the end of the outer jacket to provide a means of attaching the pulling tape. (See Figure 6–14.) The Kevlar layer is epoxied to the outer jacket and the inner jackets to prevent inducing twisting forces while the cable is being pulled, since the Kevlar is wrapped around the inner jackets in a helical pattern. The free ends of the Kevlar fibers are inserted into a loop at the end of the pulling tape and then epoxied back to themselves. The connectors are protected by foam rubber and heat-shrink sleeving. The heat-shrink sleeving is clamped in front of the steel ring in

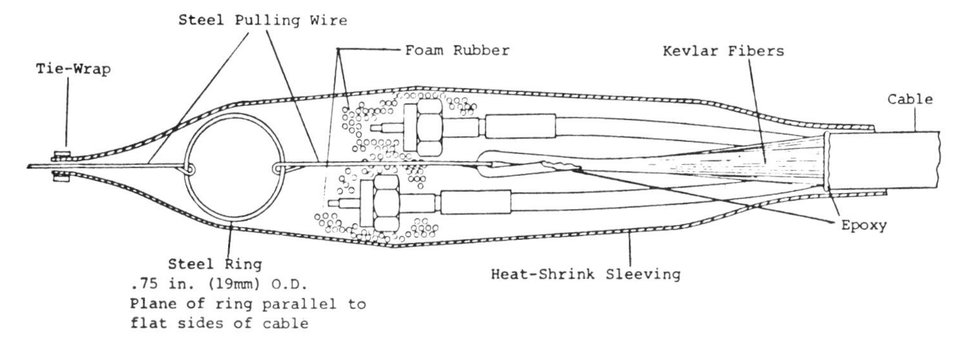

Figure 6–14 Attaching Pull Wire to Simplex Cable *(Courtesy Canoga Data Systems)*

the pulling tape to prevent pushing the connectors back toward the rest of the cable.

During pulling of the cable, pulling force should be constantly monitored using a mechanical gage. If any increase in pulling force is noticed, pulling should immediately cease and the cause for the increase be determined. If the pulling wire is subject to friction, the tensile force on the pulling wire will be more than the force applied to the fiber-optic cable, resulting in false readings. The actual tensile load on the fiber-optic cable can be measured by using a remote-reading electronic tension gage, with the sensing element located at the junction of the pulling wire and the fiber-optic cable. Optical properties of the cable can also be monitored during pulling-in using an Optical Time-Domain Reflectometer (OTDR).

During pulling-in, the cable should be continuously lubricated as it enters the duct or conduit, in the same manner as is standard procedure for wire cables. At points such as pull boxes and manholes, where the cable enters the conduit at an angle, a pulley or wheel should be used to ensure that the cable does not scrape against the end of the conduit and/or make sharp bends. If the portion of the cable passing over the wheel is under tension, the wheel should be at least 12 in. (300 mm) in diameter. If the cable is not under tension, the wheel must be at least 4 in. (100 mm) in diameter.

As the cable emerges from intermediate point pull boxes, it should be coiled in a figure-eight pattern with loops at least 1 ft in diameter. When all the cable is coiled and the next pull is to be started, the figure-eight coil can be turned over, and the cable paid out again from the top. This will eliminate twisting of the cable. The amount of cable that has to be pulled at one time can be reduced by starting the pull at a pull box as close as possible to the center of the run. Cable can then be pulled from one spool to one end of the run, then the remainder of the cable unspooled and coiled in a figure-8 pattern, and then pulled to the other end of the run. In almost all cases, the installation of fiber-optic cable is identical to that of installing conventional copper cable.

REFERENCES

Properties of optical fibers and their installation procedures were reprinted with permission of Canoga Data Systems, Canoga Park, California.

Photographs of fibers, cables, and their structures were provided by Siecor Optical Products, Inc., Horseheads, New York.

7

Fiber Testing

As with any industry product, testing of optical fiber is accomplished to ensure that the fiber qualifies under the requirements of specification. Usually, tests fall within the major headings of mechanical, optical, electrical, chemical, and physical. This does not preclude the possibility of a company to perform a special test which is peculiar to its own fiber. Tests may provide empirical or historical data which help to improve the product or production cost. Tests are performed under controlled conditions so as to obtain comparative results. The conditions in mind are temperature, humidity, and atmospheric conditions. Other conditions, such as cleanliness, and lighting, accuracy of equipment, and personnel proficiency, are considerations to be met.

FIBER TESTING

There are literally dozens of tests which can and are performed on optical fibers. Since there is not space to describe all of these, the author has chosen several mechanical and optical tests which are definable within the scope of this book.

Fiber Tensile Strength

Tensile strength of a fiber is an extremely important parameter. It allows the designer of optical fiber systems more flexibility in the choice of components in the system. Further, it provides cable manufacturers with that information which may afford the designer a margin of selection for the strength members within the cable. These are opportunities that behoove fiber engineers to develop a method of testing fiber for tensile strength.

Tensile strength of a fiber is governed by stress concentrations along the fiber. It appears from experience data that surface flaws are principally responsible for high stress concentrations. Failure usually occurs at the deepest surface flaw. Empirical data have also shown that moisture imposed on the fiber surface (and therefore the flaws) tends to enlarge the flaws and cause stress corrosion. Stress corrosion has a fatigue limit that may cause premature failure.

Accurate measurement of tensile strength is complicated by the fact that fibers are usually long. Therefore, testing involves checking of tensile strenth for selected or random lengths of the fiber. By this method, large but rare flaws may be detected and then studied to determine their causes. Once the rare flaws are known, a sophisticated system of testing may be employed.

Proof Test

Each fiber manufacture has a method of checking rated fiber tensile strength. Figure 7–1 illustrates a test setup whose principle of operation would parallel tensile strength testing in the industry.

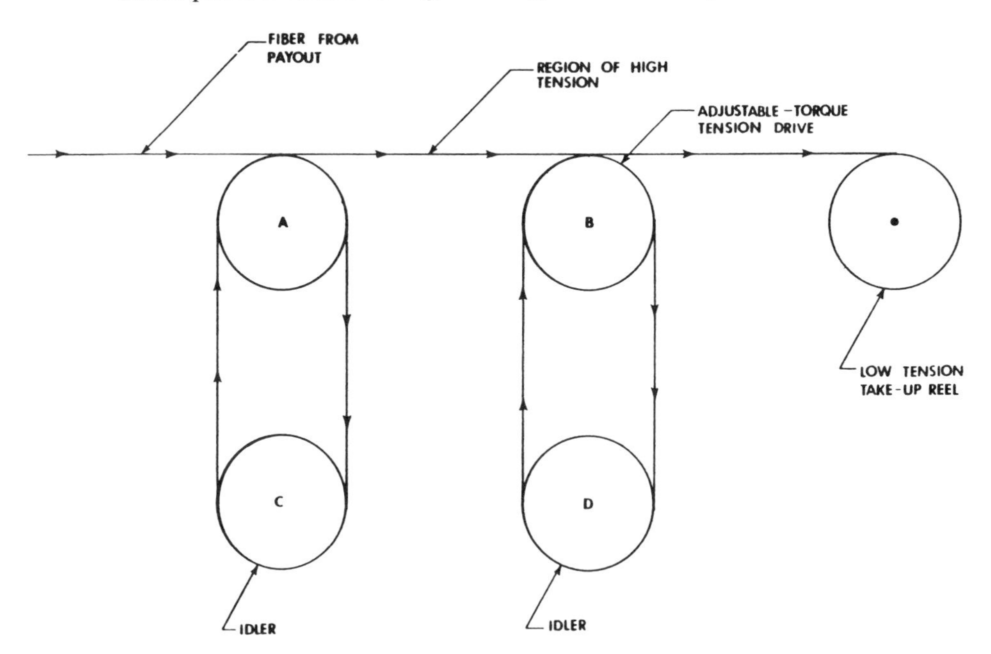

Figure 7–1 Fiber Proof Testing Test Setup *(Courtesy ITT Cannon Electric)*

An apparatus is set up with four equal-diameter pulleys (A, B, C, D). The pulleys are mounted on axles on a strong mounting structure. Belt-driven motors provide rotation.

Pulley A is motor driven at a constant speed. A magnetic clutch on pulley B provides for tensile loading and rapid disengagement. The other two pulleys, C and D, are idlers.

A fiber specimen is strung through the constant-drive pulley A and the idler pulley C and then the clutch pulley B and idler pulley D. Tension level is set by adjusting torque tension pulley B to a predetermined proof load. A friction wheel within the clutch of pulley B allows it to load release quickly.

The entire length of the fiber is fed progressively through the setup while subjected to the constant tensile load for a period of time determined by the speed of the drive pulley A and the distance between the centerlines of pulleys A and B. On the figure this area is called the region of tension. After passing over pulleys B and D the fiber is wound on a takeup reel with low tension.

Testing parameters are proof load, time under load, length of fiber under load, and number of breaks if there are any.

Dynamic Short-Length Strength Test (See Figure 7–2)

As you may recall, the proof test called for the testing of rated tension which should not break the fiber. The dynamic short-length strength test has tension applied until the fiber fractures.

In figure 7–2, a length of fiber (say 2 m) is placed between spools. The top spool is a holding spool which is fixed to a tension gage. The lower spool is a constant-speed drive consisting of a threaded shaft, drive motor, a power supply, and a pulley attached to the threaded shaft. The power supply drives the motor that turns the pulley. The pulley traverses the threaded shaft at a constant rate. The entire apparatus is attached to a rail for strength.

A predetermined tensile load is applied through the lower spool drive mechanism a specified speed until the fiber fractures. The fracture load, speed, and diameter of the fiber ends are monitored and recorded.

Sample lengths from production runs are tested to accumulate an average for statistical and historical data. Data accumulation includes average tensile strength, load rate, fiber identification, and the number of samples.

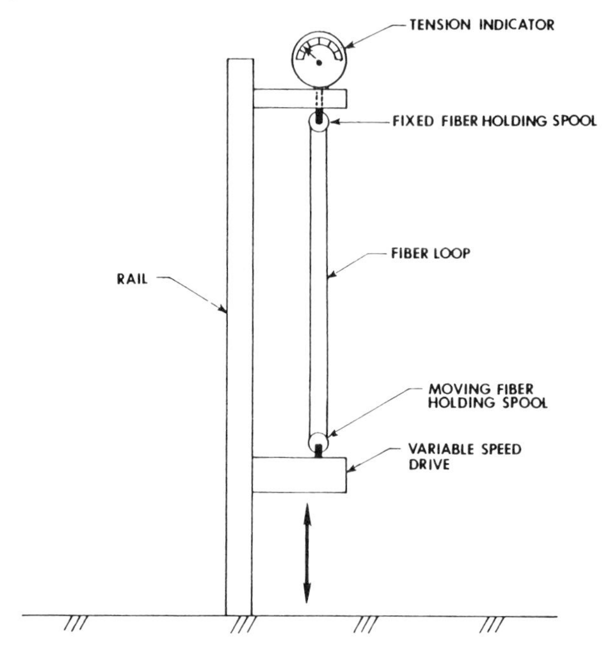

Figure 7–2 Dynamic Short-Length Strength Test Setup *(Courtesy ITT Cannon Electric)*

Static Short-Length Strength Test (See Figure 7–3)

The difference between static and dynamic tests is obvious. The fiber moves in a dynamic test and is in a still or static configuration in a static test. As in the dynamic test, loads are applied to the fiber until it fractures.

In figure 7–3 a specimen of fiber is placed between two pulleys under tension to prevent sagging. The pulleys are of equal diameter. A bracket on the right pulley in the figure is mounted to holding posts and/or tension gages. One or the other pulley has a ratchet jack for application of tension. The fiber ends are stripped of their jackets. A light source is applied to one end of the fiber and a convenient observation source on its other end. Two gage markers and a measuring tape are placed on the cable (fiber) between the two pulleys.

Tension is applied to the fiber by operating the ratchet jack. Tension is applied at predetermined increments and monitored on the tension meter. The fiber length between gage markers is measured at each increment until the fiber fractures.

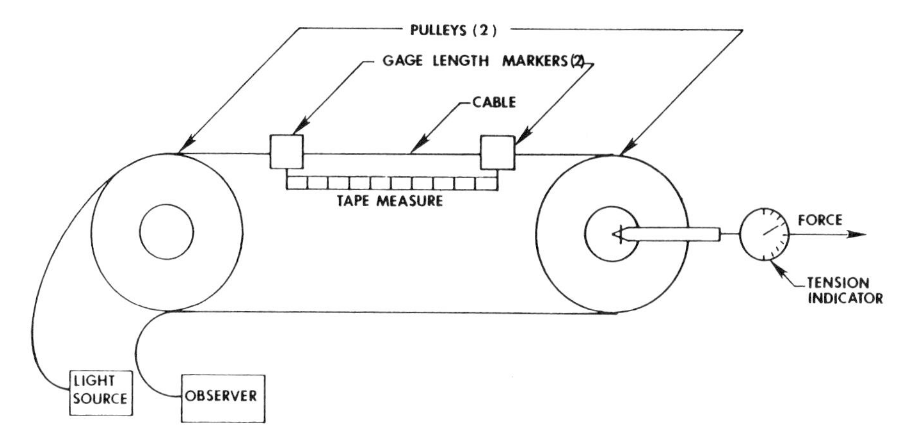

Figure 7–3 Static Short-Length Strength Test Setup *(Courtesy ITT Cannon Electric)*

Parameters that are significant are tensile load, fiber elongation, and load at fracture. Fiber load is 50% of the tension gage indication.

Some General Test Procedures Used In the Fiber-Optics Industry

Test procedures are varied and range from simple to complex. The test procedures listed here are just a small number of ones that are in actual use. Each manufacture will also have their own method or technique to perform the test for best results and economical speeds of accomplishment.

Fiber Size Measurement This test measures the core and cladding of an optical fiber. The test is accomplished using a microscope, a prepared fiber, and a predetermined calibration index.

Fiber Bundle Measurement The measurement checks bundle size by dipping the bundle into wetting liquid, then placing the bundle into sizing gages beginning with the largest gage.

Number of Fibers Determination of the total number of fibers in a fiber bundle where the number of fibers is large. The fiber bundle end is prepared and photographed. The photo is enlarged to applicable size and the fibers are counted.

Number of Transmitting Fibers Test is used to determine the number of fibers in a bundle which are transmitting light. Several photographs are taken using different intensity settings. Photos are enlarged as necessary and the fibers transmitting are counted.

Cyclic Flexing A procedure used to determine the ability of a fiber-optic cable to withstand cyclic flexing. The fiber is placed in a cyclic arm driven by an electric motor. The cyclic arm flexes the fiber a specified count.

Low-Temperature Flexibility A test which determines the ability of a fiber optic cable to withstand bending around an obstacle at low temperature by measuring fiber breakage or transmitted power. The fiber is bent around mandrels, then straightened while under cold temperatures.

Impact Testing A procedure used to determine the ability of a fiber-optic cable to withstand impact loads. A drop hammer of specified weight free falls on a length of fiber during this test to determine damage.

Compressive Strength A test to check the ability of a fiber-optic cable to withstand slow compression or crushing. A fixture which applies compression to the fiber is used and an inspection is made to determine damage.

Cable Twist and Bend Test is used to determine the ability of a fiber-optic cable to withstand twisting. Checks are made for broken fibers and/or attenuation. A pair of gripping blocks hold a length of fiber. One of the blocks is fitted with a rotating clamp to twist the fiber. Fiber is then inspected for damage.

Cable Tension Load This test is used to determine percent increase in length under tension, broken fibers, changes in radiant power, and the ability of a strength member to withstand tensile loads. Load is placed on the fiber in a form such as gripping blocks. The fiber is placed under tension. Measuring gages check tension and length.

Power Transmission Versus Temperature The effects of temperature on the transmitted power of an optical cable are defined by this test. This is a relation between two temperatures. A light source and detector are placed on either side of a test chamber. Transmission is made while temperature is varied.

Power Transmission Versus Temperature Cycling A test to measure the effects of temperature upon the transmitted power of an optical cable while cycling two temperatures. Fiber is placed in high, then low, test chambers. The light source and detector are on the outside of the test chamber. The fiber is exposed in cycles to high and low temperatures.

Power Transmission Versus Humidity This test determines the effect of humidity upon transmitted power. This is a ratio of radiant power of initial humidity preconditioning and at a final condition. A fiber specimen is placed in a humidity-controlled test chamber with the light source and detector externally mounted. Control is made to ensure that condensation does not fall on the fiber.

Tensile Loading Versus Humidity This test defines the effect of humidity on the tensile loading of a fiber-optic cable. Test is performed within a test chamber. Loads are placed on the fiber under controlled humidity.

Freezing Water Immersion—Ice Crush A test to determine the effect of crush force caused by freezing water on the transmitted power of an optic cable immersed in freezing water. This is a ratio of radiant power in cold but unfrozen water and in frozen water. Fiber is placed in a test chamber which is able to lower temperatures quickly at a predetermined rate.

Dimensional Stability This test determines the permanent dimensional changes which occur to a fiber jacket or covering when the cable is exposed to elevated temperatures. The test checks expansion and shrinkage usually in a percent of change. The test fiber is dimensionally checked before and after exposure to heat in a test chamber.

Flammability The test measures the flammability of a subject fiber by application of a flame such as a bunsen burner, then checking for fiber damage.

Far-End Crosstalk The test measures the crosstalk of two transmitting fibers which are neighbors. Two fiber specimens are placed parallel to each other, radiation is applied to one fiber, and detectors are used to monitor radiation emanating from the illuminated and neighboring fibers.

Refractive Index Profile This test determines the refractive index profile of a graded-index fiber by the interferometric process. The fiber end is prepared and placed under a microscope. The microscope is tilted so as to view a tilted wavefront on the fiber face. A photograph is taken for empirical data.

Military Specifications and Standards

Test procedures for fiber optics fall into many categories; the main ones are general physical, electrical, and chemical methods. Most of these standards and specifications are referenced to federal and military documents.

Specifications

Federal

L-P-390	Plastic Molding Material, Polyethylene, Low and Medium Density
QQ-W-343	Wire, Electrical and Nonelectrical, Copper (Uninsulated)
QQ-W-423	Wire, Steel, Corrosion-Resisting

Military

MIL-C-17	Cables, Radio Frequency; Coaxial, Dual Coaxial, Twin Conductor and Twin Lead
MIL-I-631	Insulation, Electrical, Synthetic-Resin Composition, Nonrigid
MIL-W-5086	Wire, Electric, Hookup and Interconnecting, Polyvinylchloride-Insulated, Copper or Copper Alloy Conductor
MIL-C-12000	Cable, Cord, and Wire, Electric, Packaging of
MIL-C-13777F	Cable, Special Purpose, Electrical
MIL-M-20693	Molded Plastic, Polyamide (Nylon) Rigid

Standards

Federal

FED-STD-191	Textile Test Methods
FED-STD-228	Cable and Wire, Insulated; Methods of Testing
FED-STD-601	Rubber, Sampling and Testing

Military

MIL-STD-104	Limits for Electrical Insulation Color
MIL-STD-109	Quality Assurance Terms and Definitions
MIL-STD-202	Test Methods for Electrical and Electronic Components
MIL-STD-810	Environmental Test Methods

8

Coupling

Optimum alignment must be made each time an optical fiber is tied together (spliced), coupled to a connector, and presented to a source and/or a detector. Clean ends are necessary for all the above. Ends joined in lengthy sections are included. In fact, the science of coupling has become the major problem and the significant attribute of the fiber-optic industry. For it is here where power losses are either destructive or an enhancing attribute.

FIBER END PREPARATION

Before placing a fiber into a connector, it must be properly prepared. The jacket of the fiber must be removed; the fiber is then broken and the fiber end is ground and polished. Procedures involve cleaning and coating the fiber with a suitable refractive index matching compound. Final procedures also include, of course, thorough inspection.

INSPECTION CRITERIA (See Figure 8–1)

After polishing and cleaning, the terminated ferrule end should be inspected. Acceptable terminations should look somewhat as the two photographs on the left side of figure 8–1. Note the mirrorlike finish. The fiber should have no cracks in its core. The surface of the fiber should be flush with the face of the jewel or slightly concave. Under no circumstance can the fiber be convex.

The photographs in the right side of figure 8–1 illustrate unacceptable terminations. As shown, these fibers are under 400× magnification.

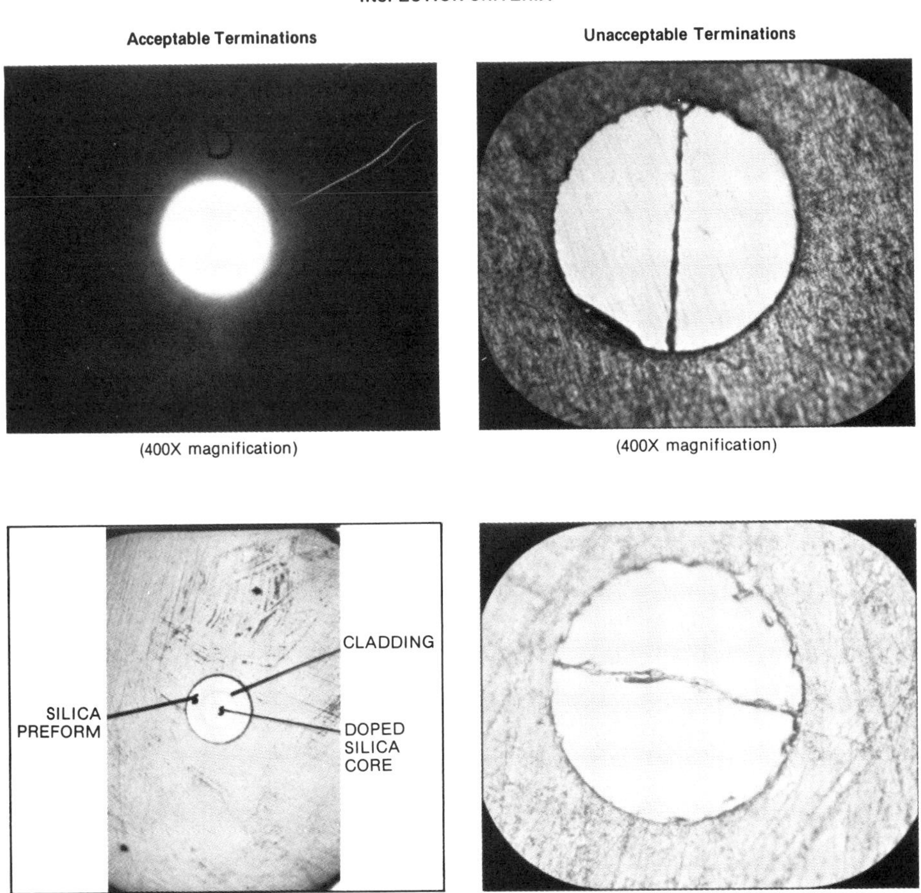

Figure 8–1 Inspection Criteria *(Courtesy ITT Cannon Electric)*

SPLICING

There are dozens of different splices and splicing techniques used in the fiber-optic industry. All splices have several aspects in common. They all must have the attribute of low power loss. Attenuation must be held to as near 0.1 dB as possible. Splices must be quick to install. All should be lightweight, extremely strong, and small.

Probably the most critical one requirement of splicing is alignment of the fiber. Alignment philosophy and mechanics were covered in detail in previous paragraphs.

Although fiber-optic manufacturers provide a large variety of splices, most of these utilize several basic techniques.

The Fused Splice

The most difficult splice to make is the fused splice (see figure 8–2). It is also the splicing technique that provides the lowest attenuation. Economics and efficiency are a trade-off. Two fiber ends are placed in a tooling fixture which aligns the ends to be fused. Fiber ends are polished and prepared. The fibers usually are set in V-grooves in the fixture and are clamped in resilient vise jaws. Heating elements are then brought to the connecting fiber ends. Heat is applied and the fusing takes place. Several fusion heaters used are micro torches, Nichrome wire heaters, and electric arc fusers.

The Siecor Model 65 Fusion Splicer

This equipment employs the recognized method of permanently fusing two butted optical fibers together by heating with a tiny, well-regulated electric arc. (See figure 8–3.) The field-usable splicer consists of fiber-holding and alignment devices, a viewing microscope, arc positioning and intensity controls, a built-in arc power supply (which requires only a 12-V dc

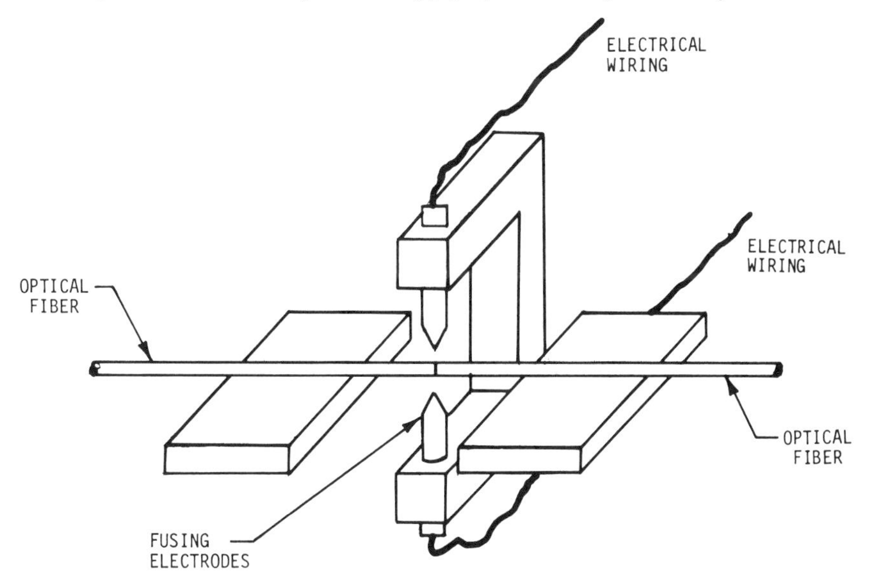

Figure 8–2 The Fused Splice

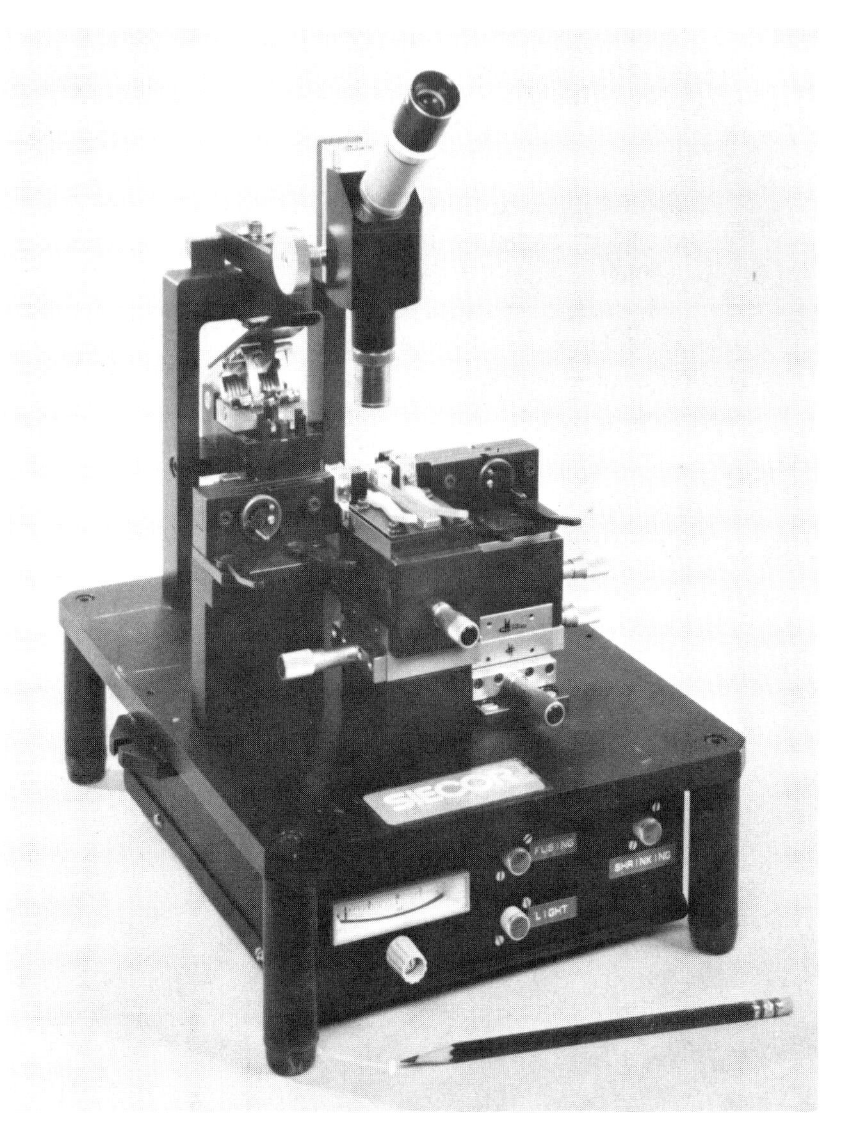

Figure 8–3 The Siecor Model 65 Fusion Splicer *(Courtesy Siecor Optical Cables Inc.)*

source), and a fiber-cutting tool. Typical splice losses between telecommunications-grade optical fibers which are achieved with this device are from 0.2 to 0.4 dB.

Field experience has shown that the Model 65 requires no extensive operator training or experience for the production of low-loss splices. After the fibers are cleaned and cut, they are clamped in the holders and positioned with the aid of the microscope. The arc is used briefly to burn off the fiber coating and then to slightly round the fibers' ends. With a higher arc power, the fibers are fused together. Monitoring splice loss during the actual operation is desirable but not necessary. After splicing, the joint is recoated and placed in a protective sleeve before going into the splice closure.

The Precision Tube Splice

A second method of splicing is the precision tube method (see figure 8–4).

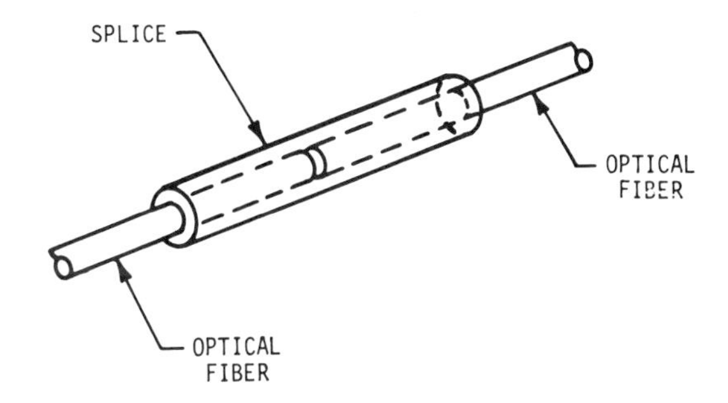

Figure 8–4 The Precision Tube Splice

This method utilizes a precision tube that has been machined to a close tolerance and will accept the fiber diameter. Fiber ends are polished and prepared. A splicing compound may be injected into the splice. The splicing compound has the same refractive index as the fiber to be spliced. The fiber ends are inserted into the splice. A precision guide may be used to accommodate the insertion. The fiber ends are butted together within the splice. The outer jacket of the fiber is crimped. Care must be taken during crimping not to disturb the cladding and core interface. Crimping tools suggested by the fiber manufacture are an absolute must. In some cases, the fiber splice is surrounded by a metal case or sleeve. The sleeve is crimped so as not to disturb the critical makeup of the fiber. Thermal shrink tubing is an acceptable alternative to crimping.

The Loose Tube Splice

Still another splicing method, called a loose tube splice, uses a rectangular tube for splicing (see figure 8–5). The tube has a square hole which is curved slightly within the splice. Fiber ends are polished and prepared. A

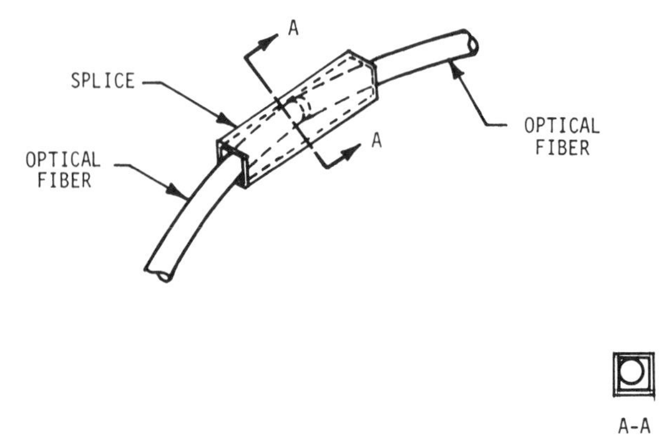

Figure 8–5 The Loose Tube Splice

splicing epoxy is injected into the splice. The epoxy has the same refractive index as the fiber to be spliced. The fiber ends are inserted into the splice. The fiber ends must be slightly bent to insert. The fiber ends are butted together. Rotation of the square tubes guides the fiber ends together in the V-groove in one corner of the splice. Once the alignment takes place, the epoxy is allowed to set (cure).

The Double Metal Plate Splice

A splicing method that is widely used in some form or another is the double metal plate splice (see figure 8–6). Also called the grooved substrate splice, this technique involves two machined metal plates. The plates have a V-groove down their center which has been geometrically matched to the size of the fiber to be spliced. A copper or silicon substrate may also be coated on the metal backing and etched to ensure the correct V-groove to match the fiber diameter. Fiber ends are polished and prepared. A splicing epoxy is placed in the V-grooves on each metal plate and on the fiber ends. The epoxy must be refractive index matched to the fiber. The fiber ends are placed on one machined plate and the ends are butted together. The second machined plate is placed over the V-groove, aligned and installed on the first plate. Suitable methods of fastening the two metal plates together are applicable.

The V-Shaped Splice

The Siecor Model 60 Mechanical Splicer is a small, portable device (see figure 8–7) which can connect the fibers of two or more optical cables in a

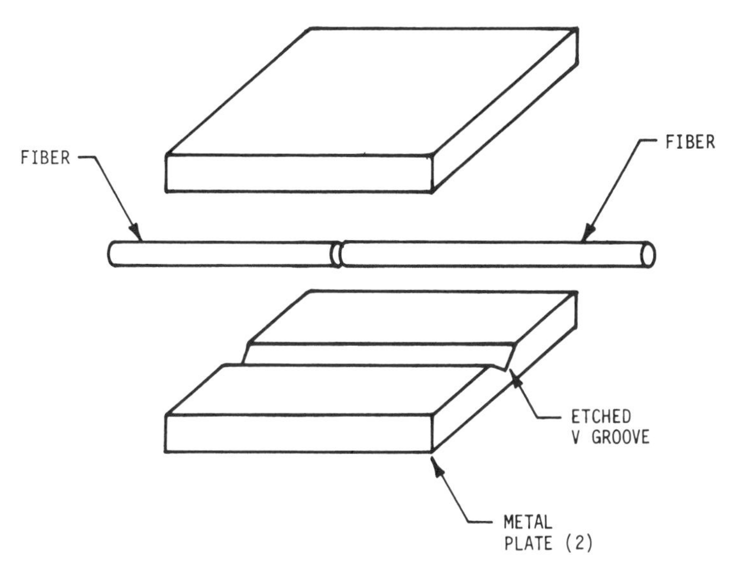

Figure 8–6 The Metal Plate Splice

Figure 8–7 The Siecor Model 60 Mechanical Splicer *(Courtesy Siecor Optical Cables Inc.)*

simple, reliable manner with no need for touching or handling the fiber. It consists of two mounted swiveling clamp assemblies to hold the buffer jackets and fibers, a sliding tool to scribe and break the fibers to a precise, preset length, a cam-operated fixture for holding and positioning a V-groove splice part, and means for holding a splice case internal mounting plate.

All components are mounted on a base plate. Also incorporated in the splicer is a fixture to hold the splice closure's internal mounting plate which greatly simplifies the procedure of making individual fiber splices with multifiber cables. Figure 8–8 illustrates the splicing sequence.

The Siecor technique is based on the principle of centering the fibers to be joined on the bottom of a V-shaped alignment part after breaking the fibers normal to their axes and at the exact required length. A magnetic support holds the V-groove alignment part and is cam-actuated to lift it to the exact alignment position. By raising this V-shaped splice part upward,

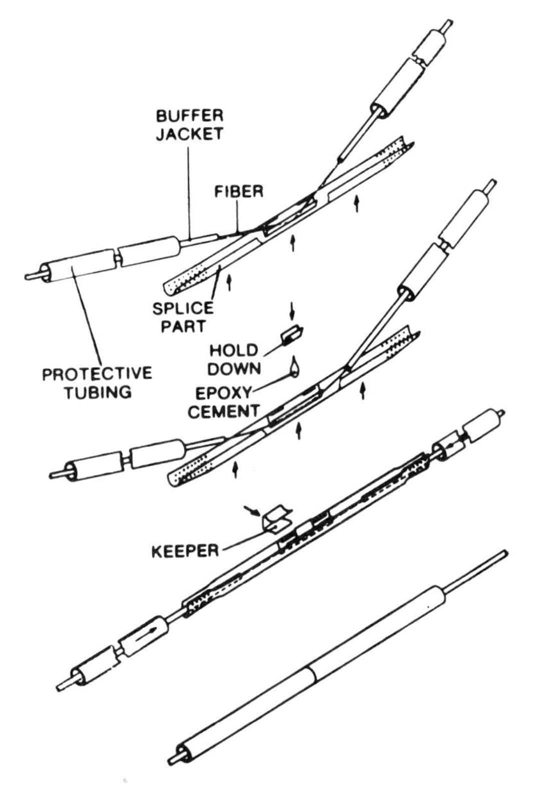

Figure 8–8 Splice Sequence Using the Siecor Model 60 Mechanical Splicer *(Courtesy Siecor Optical Cables Inc.)*

the fiber ends, as a result of fiber stiffness, are pressed together and are aligned themselves on the bottom of the V-groove.

A droplet of transparent, two-component epoxy holds the fiber ends in position, acting as an index matching substance at the same time. After the epoxy has cured, the buffer jackets of the fibers spliced together are clamped by the serrated ends of the splice parts.

There are, as you may have deduced, many other splicing techniques. Roller bearing rods externally surrounding a fiber splice and held together with shrink tubing are used. Precision-molded plastic machined metal, or etched silicon, have been extensively involved with fiber bundles. Proprietary splices will always be a significant part of the market. Until industry can standardize splicing and connecting fiber ends, the intricacies of the techniques involved will be an economical marketing entity.

CABLE CONNECTORS

Acceptable cable structures must achieve thermal and elastic compatibility, and stress fibers to a controlled and uniform degree under both tensile and crush loading. They must prevent fiber-to-fiber contact, especially in a crossover configuration. In-line system connectors usually fall into two specific categories. The first has the fiber group butted collectively against the second group of fibers. The second butts individual fibers. In group connections losses in transmission are typically 3 to 5 dB. The group connectors are easily made. The single-fiber connection has relatively low losses, in the neighborhood of 0.5 dB, but are more expensive.

The desirable properties of waveguide connectors are limited by the bundle-to-bundle connectors. In addition to the severe insertion loss, grinding and polishing are required, and epoxy or other glue limits the connector performance, especially with regard to temperature, humidity, and chemical attack.

Individual fiber-to-fiber connectors will substantially eliminate the objections to the bundle-to-bundle connectors. However, well-matched fibers and precision connector hardware are required. In addition, care must be taken not to distort the fibers, which can cause significant scattering loss. This effect is encountered especially in connectors using adhesives (epoxies), due to material inhomogeneity or nonuniform curing. Finally, fiber ends must be smooth to avoid another cause of scattering losses. Fiber-end losses can be minimized through the use of index-matching fluids, which also reduce Fresnel reflection losses.

In addition to the need to minimize insertion loss in a structure mechanically and environmentally suited for its use and compatible with the cable, a connector should not require adhesive to hold the fiber, or grinding and polishing to finish the fiber end. Simplicity of connector hardware and of the termination process is important relative to future production costs and indispensable for practical field termination.

One connector concept locates the waveguide in the interstices of appropriately dimensioned rods of elastically suitable materials. (See figure 8–9.) It is immediately apparent that this approach has a self-aligning capability minimizing requirements for tight fiber tolerances. The rods are made of a compliant material for fiber retention without the use of adhesives. Due to crosstalk considerations, only every other interstitial space is used.

A second cable-connector concept is the overlap concept. The connector assembly technique allows simultaneous cutting of all waveguides based on controlled scoring of bent and stressed fibers. This technique has consistently viable perpendicular and fault-free fiber ends, eliminating the need for grinding and polishing.

A viable connector concept must not impose the optical dimensional tolerances on the mechanical elements of the assembly. This is accomplished by separation of the mechanical and optical mating planes as shown in figures 8–10 and 8–11. In addition, the waveguides are made to function as springs, assuring contact of fiber ends. This is likely to prove particularly important under shock and vibration conditions over a broad operating temperature range.

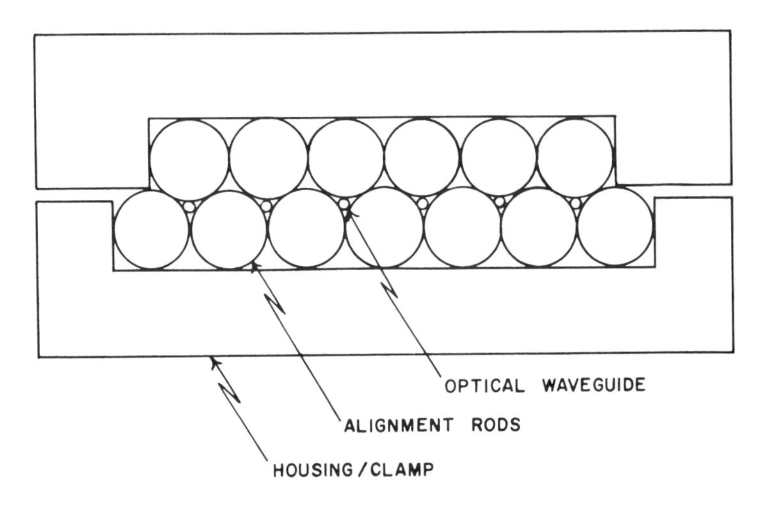

Figure 8–9 Typical Cable Connector Concept *(Courtesy Corning Glass Works)*

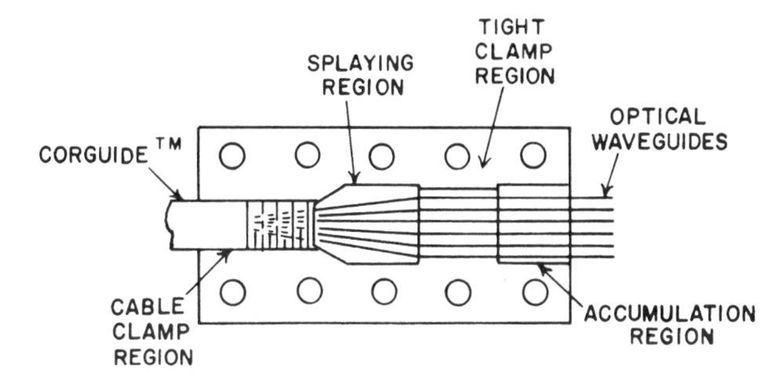

Figure 8–10 Integration of Cable End into the Overlap Connector *(Courtesy Corning Glass Works)*

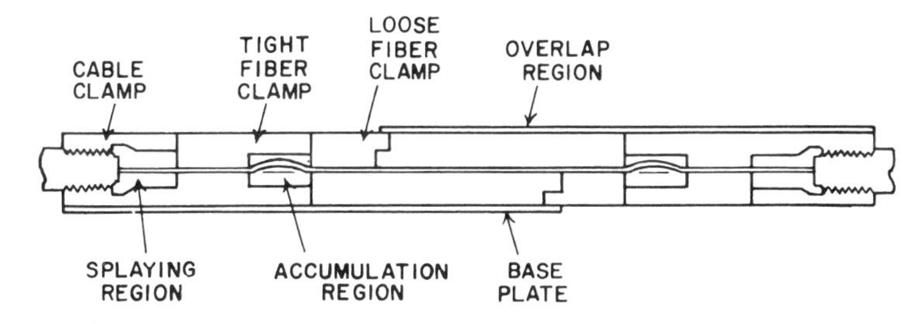

Figure 8–11 The Overlap Connector *(Courtesy Corning Glass Works)*

A TYPICAL CONNECTOR INSTALLATION

One of the many connector installations used in the fiber-optic discipline is the Deutsch connector and termination method. The primary advantage of using the Deutsch method is that there are no requirements for epoxy, grinding, and polishing. Secondary advantages are minimum training requirements and the capability of easy field use in a short time.

The procedures are written around the use of the fiber-optic termination tool shown in figure 8–12. The tool, of course, uses the score and break process. This may be compared favorably with the wet process system of terminating. Probably more important than any other attribute is the procedure provides precise alignment with less than 1-dB insertion loss.

Procedures that follow install the plug on the cable, then terminate the fiber.

Figure 8–12 The DW9000 Fiber Optic Termination Tool *(Courtesy Deutsch Co.)*

Installation of Cable on Plug

Procedures are as follows. Illustrations are shown in figure 8–13A through H.

1. Slide the cable nut over the cable and down several (12) inches from the end of the cable. Arrange the washer, O-ring, sleeve, and sleeve insert along the cable assembly. (See figure 8–13A.)
2. Use the cable stripper, utilizing blades 10, 14, and 22 to strip the cable layers below the point of the sleeve insert. Leave about ½ to 1 inch of Kevlar—enough so that it will fold back over the rear of the sleeve insert. (See figure 8–13B.)
3. Slide the Kevlar retaining tool DW9001 over the Kevlar to hold it back for installation of the sleeve. (See figure 8–13C.)
4. Slide the sleeve over the Kevlar, which is folded back over the sleeve insert. Remove the tool and slide the sleeve over until it covers the sleeve insert. (See figure 8–13D.)
5. Slide the O-ring up over the sleeve until it snaps into place in the groove at the rear of the sleeve. (See figure 8–13E.)
6. Slide the washer up to the rear edge of the sleeve. (See figure 8–13E.)
7. With one hand, hold the clamp nut while the other hand pulls the cable back through the clamp nut so that the sleeve assembly is pulled to the rear of the clamp nut, placing the cable under tension. (See figure 8–13F.)

8. Insert the wedge sleeve until it is flush with the sleeve assembly within the backshell. (See figure 8–13G.)
9. Screw the backshell nut onto the plug while maintaining tension on the cable. (See figure 8–13G.)
10. Follow the termination procedure shown in figure 8–14. If the termination is not satisfactory, repeat the procedure. (See figure 8–13H.)

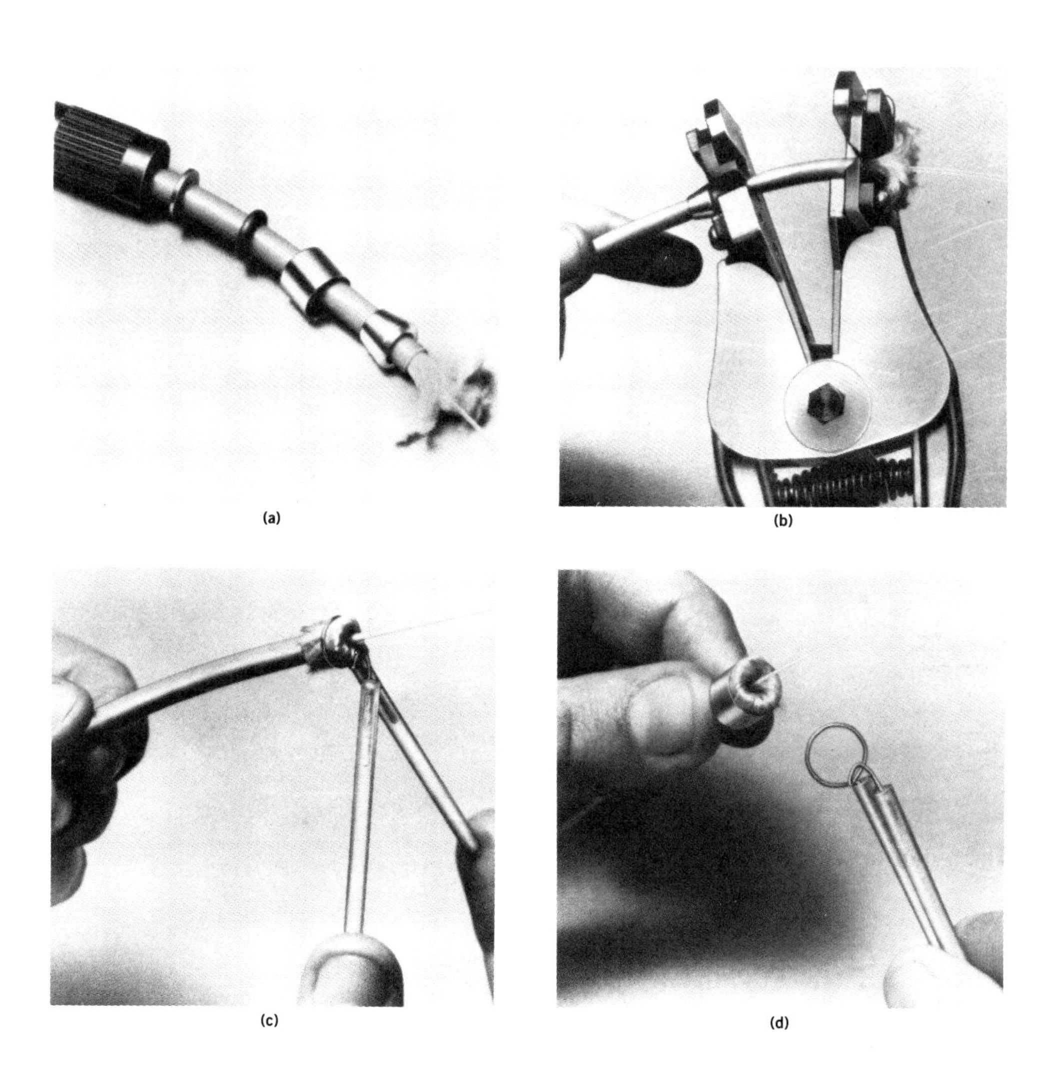

Figure 8–13 Optical Waveguide Connector Plug Assembly *(Courtesy Deutsch Co.)*

Figure 8–13 *continued*

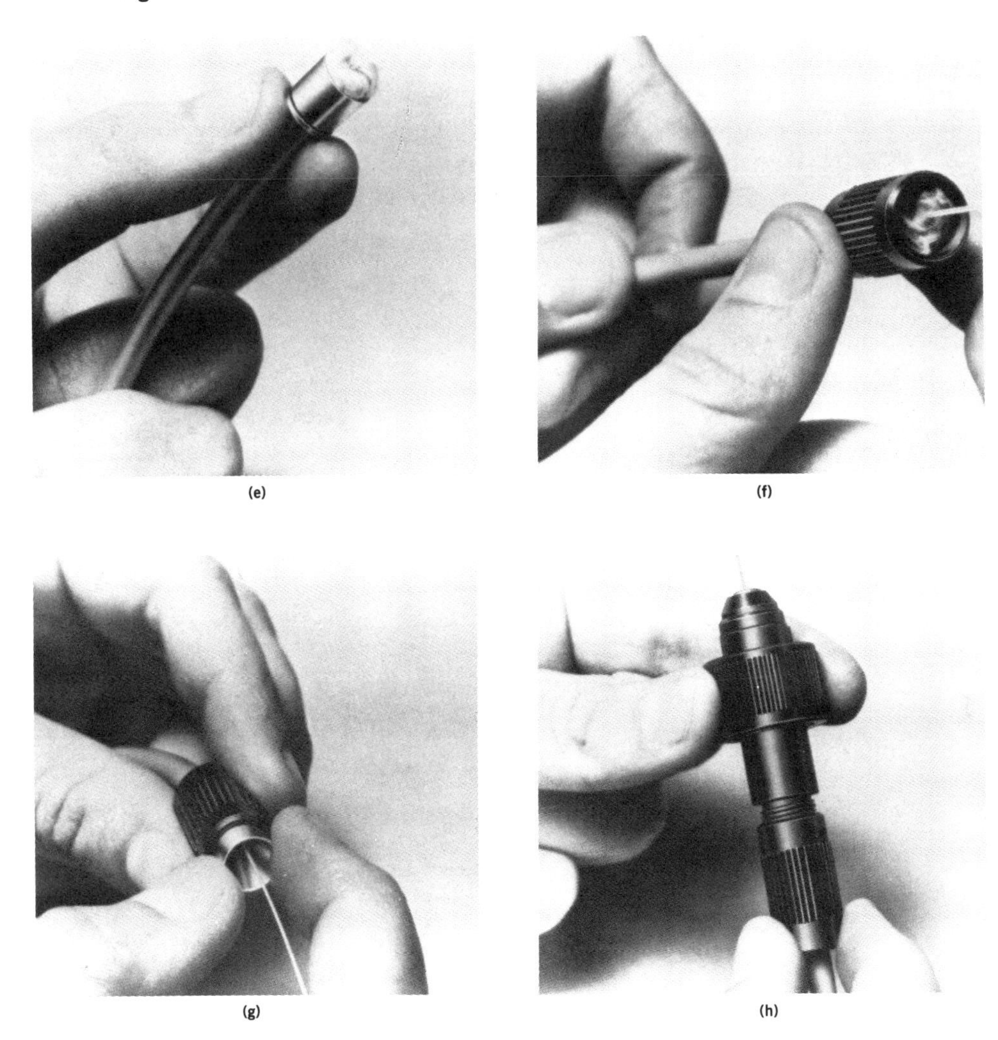

(e) (f) (g) (h)

Termination Procedures

Procedures are as follows. Illustrations are shown in figure 8–14A through G.

1. Loosen the backshell of the optical plug until the fiber can be inserted easily through the plug. (See figure 8–14A.)
2. Carefully insert the fiber through the backshell and plug so that at

least 1 inch of fiber protrudes from the front of the plug. (See figure 8–14B.)

3. Tighten the backshell until the fiber is held secure. (See figure 8–14C.)
4. Align the key on the plug with the keyway on the tool and insert the plug into the plug retainer. Tighten the coupling ring until completely inserted. (See figure 8–14D.)
5. Squeeze the handles together and release. Upon the release of handles, the scoring blade will be activated to score the fiber and make an optical-quality break. (See figure 8–14E.)
6. Remove the plug from the plug retainer. (See figure 8–14F.)
7. Remove excess fiber from the tool. (See figure 8–14G.)

Note: The reader must realize that connectors and cables available have their own peculiarities. It is imperative that the user consult a company such as Deutsch Corporation to ensure that correct procedures are used. The steps written here are for example only. It must also be realized that new and improved connectors along with cable termination procedures are being produced daily. State-of-the-art devices and tooling are short-lived in this rapidly advancing field of electronics.

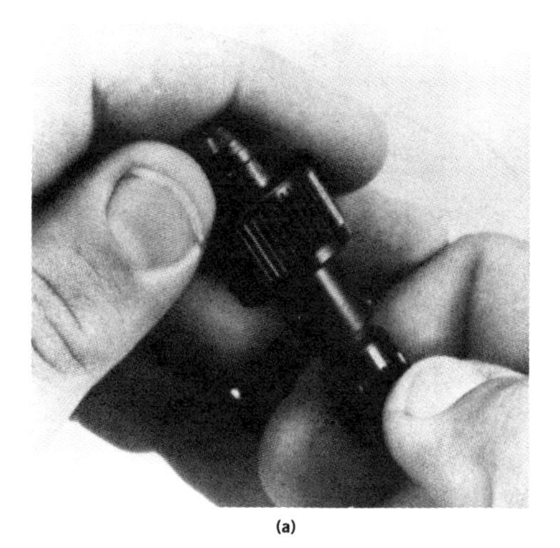

(a)

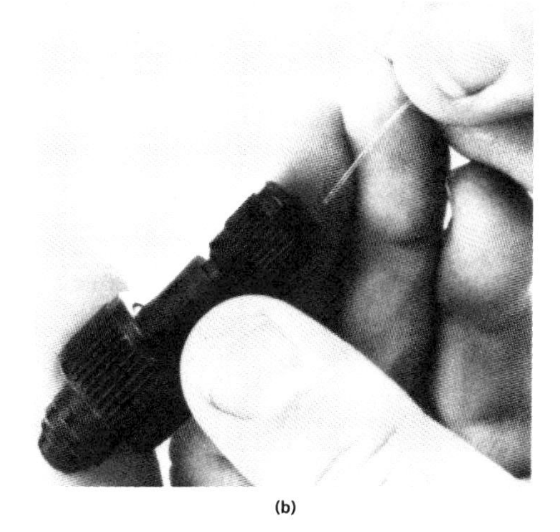

(b)

Figure 8–14 Optical Waveguide Termination Procedures *(Courtesy Deutsch Co.)*

Figure 8–14 *continued*

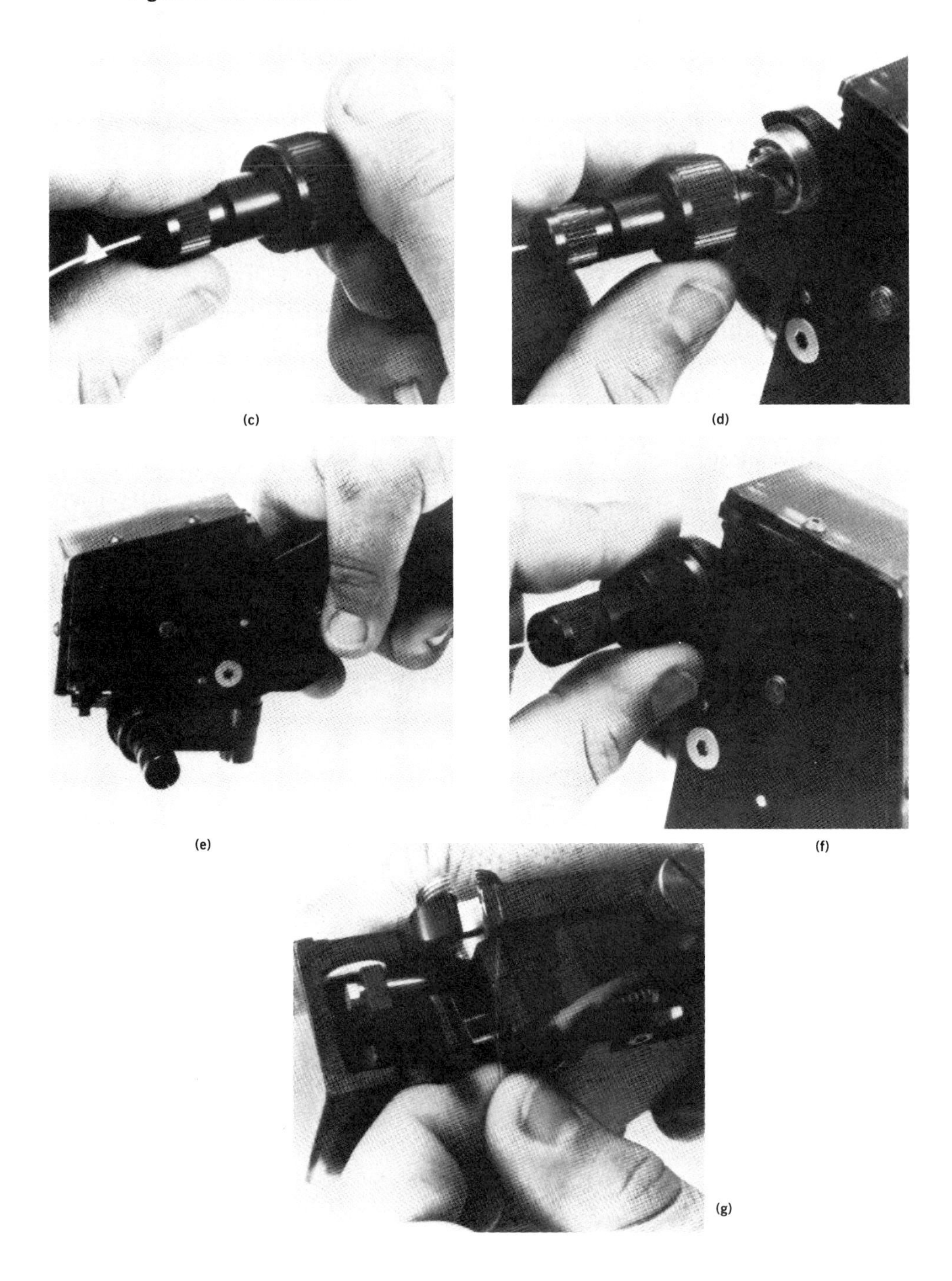

(c) (d) (e) (f) (g)

9

Light Sources and Detectors

The purpose of the light source is to launch a light signal into the optical fiber at an angle that provides maximum signal transfer. The purpose of the detector in a fiber-optic system is to detect (receive) the light signals as they are launched by the light source.

THE LIGHT SOURCE

There are two basic light sources used in fiber-optic electronics. These are the light-emitting diode (LED) and the injection laser diode (ILD). Both of these units provide small size, brightness, low drive voltage, and are able to emit signals at desired wavelengths. Each has characteristics that make them desirable or undesirable for a particular application. The LED has a longer life span, greater stability, wider temperature range, and much lower cost. The ILD is capable of producing as much as 10 dB more power output than the LED. It can launch the light signal at a much narrower numerical aperture (NA), and therefore can couple more power through the optic fiber than the LED. The disadvantage of using an ILD is that its current range is extremely restricted. Since some system operations vary greatly, the ILD must have compensation devices added to the electronics. This may make the cost prohibitive.

THE LIGHT-EMITTING DIODE (LED)

The light-emitting diode (LED) represents the best of the electroluminescent devices. Electroluminescence, you may recall, is the emission of light from a solid-state device by application of current flow through the device. The light comes from photons of energy caused by hole-electron combinations. The diode is forward-biased from an external source. Electrons are injected into the N-type solid-state material. Holes are injected into the P-type solid-state material. The injected electrons and holes recombine with majority carriers near the PN junction. The result is radiation in the form of photons in all directions and specifically from the top surface.

Let's consider some of the materials from which the light-emitting diodes are made. Most are made from so-called compound semiconductors such as gallium arsenide phosphide (GaAsP), gallium phosphide (GaP), and gallium aluminum arsenide (GaAlAs). Gallium arsenide (GaAs) is usually included in this group but it must be understood that GaAs emits only infrared radiation around 900-nanometer wavelengths. These cannot be seen, of course, and are not generally classified as LEDs. Some manufacturers call them infrared LEDs, while others name them infrared emitters.

The GaP emits green light from 520- to 570-nanometer wavelengths with its peak around 550 nm. It can also emit red light between 630 and 790 nm with a peak of 690 nm. The GaAsP emits light over an orange-red range depending on the amount of GaP in the material. The GaAsP emits red light between 640 and 700 nm with a peak at 660 nm. With the correct amount of GaP in the material a yellow light is emitted with a peak around 610 nm wavelength. The GaAlAs emits light over a red range from 650 to 700 nm with a peak around 670 nm. The efficiency of the LED is very dependent on the emitted wavelength, with drastic falloff in efficiency as the wavelength gets shorter.

Figure 9–1 is a cross section of a typical LED used in fiber optics. To

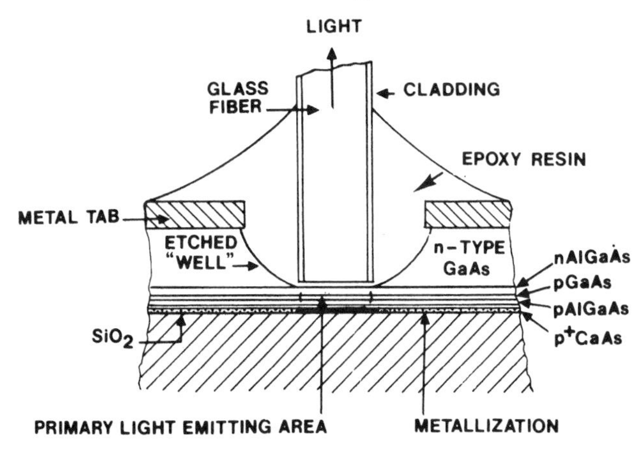

Figure 9–1 Cross Section of a Fiber-Optic LED

be useful the LED used for fiber optics must have high radiance and a very fast response. The LED is also most useful in that it can easily be directly modulated by an analog signal. Digital modulation requires that the LED be driven harder; therefore, there are more current and possible heat problems.

The characteristics most generally considered when choosing an LED as a light source for fiber optics are:

1. Wavelength
2. Spectral width
3. Power
4. Coupling
5. Current-voltage *(IV)* characteristics

LEDs used for fiber optics operate in the 810- to 850-nm wavelength range. Some new and research LEDs are being developed to use the 1100- to 1400-nm wavelength range.

Typical of these LEDs is the ITT T810 to T813 series shown in figure 9–2. Some of the series specifications are as follows:

Peak Emission Wavelength	840 nm
Spectral Width	40 nm
Peak Forward Voltage	1.5 V
Reverse Voltage	1.6 V

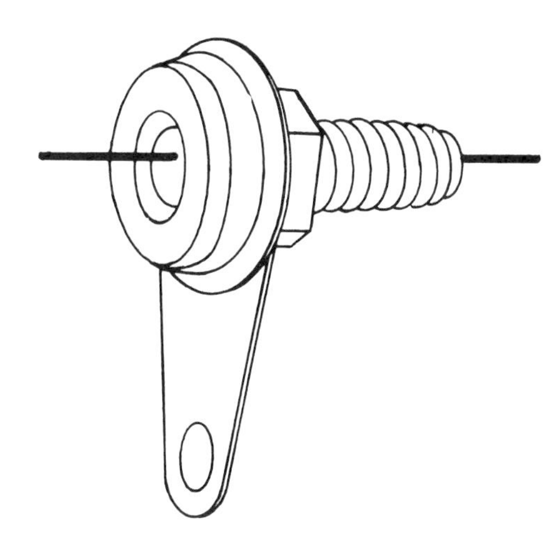

Figure 9–2 Typical High-Radiance LED Used in Fiber-Optic Systems

Continuous Forward Current	100 mA
Pulsed Forward Current	150 mA
Operating Temperature Range	−40 to 75°C
Optical Power Varies from 95 to 520 μw Depending on the Model	
Emitting Diameter	50 μm

This series of LEDs is of GaAlAs double heterostructure design. They may be used with step-index or graded-index fibers.

INJECTION LASER DIODE (ILD)

The semiconductor injection laser diode (ILD) is extremely well suited for use within the fiber-optic industry. The prime reasons for their applicability are their inherent ruggedness, extreme efficiency, and small size. They can be pumped and modulated by injection current. Figure 9–3 is an ITT Model T912 double-heterojunction AlGaAs injection laser typical of the present-day state-of-the-art ILDs on the market. The term *heterojunction* requires some explanation. Early in the development of semiconductors, lasing action by stimulated recombination of carriers injected across a PN junction was predicted. In 1962, lasing action was achieved in a crystal of gallium arsenide (GaAs) (see figure 9–4). It was accomplished by three independent research teams simultaneously. The firms were GE, IBM, and Lincoln Laboratory. The unit was called a homostructure PN-junction laser. The word *homostructure* is used because the injection laser diode was made from one material, gallium arsenide (GaAs). The lasers are called junction laser diodes, junction lasers, and injection laser diodes because electrons are injected into the junction region. The gallium arsenide (GaAs) is lightly doped with suitable impurities to form a PN junction just as silicon is doped when manufacturing solid-state diodes for use as rectifiers. The ends of the GaAs crystal are polished to a mirrorlike finish. Light is generated by injecting current (electrons) into the P-type region, where there are de-

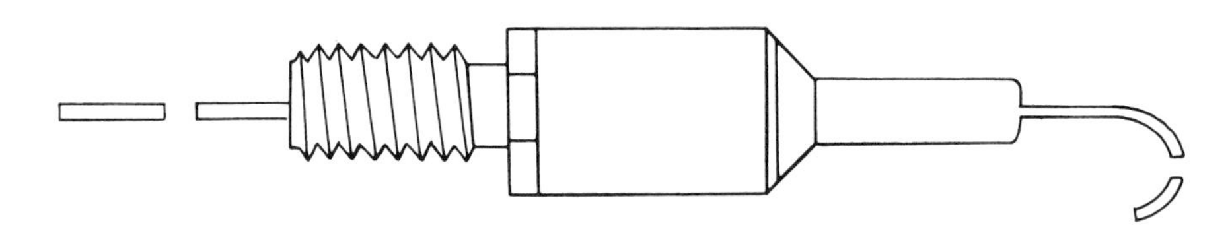

Figure 9–3 Typical Injection Laser Diode Used in Fiber-Optic Systems

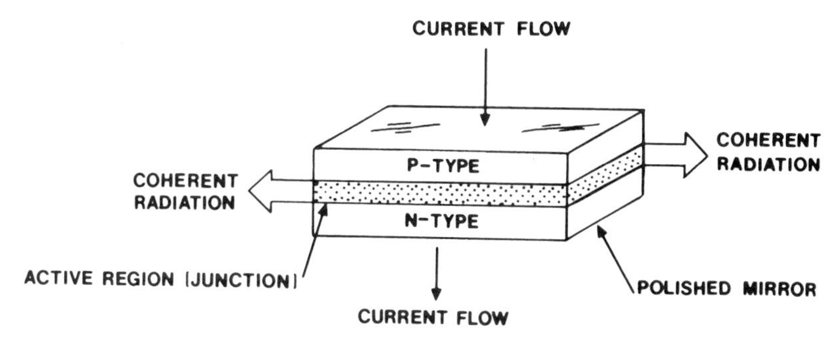

Figure 9–4 Homostructure PN-Junction Laser

ficiencies of free electrons in its lattice structure. The N-type region has an excess of free electrons. When the electrons and holes recombine within the junction region (active region), photons are emitted as radiant energy. The mirror ends tend to reflect the photons back into the active (more recombinations) and serve as a feedback mechanism. The sides of the device are optically diffused. Brightness is controlled by adjusting the current flow. As noted, the ILD is placed in a circuit in reverse of a standard diode because current (electrons) are injected into the P-type material.

The homostructure diode quickly fell into obscurity for it was found that the ILD could be improved considerably with the use of the heterostructure. In the heterojunction, the active region is bounded by wider band-gap regions. Heterojunctions fall into three categories, the single heterostructure (SH), the double heterostructure (DH), and the separate confinement heterostructure (SCH). The single heterostructure has only one heterojunction, so that injected carriers are confined by the junction at only one boundary of the active area. The double heterostructure (DH) laser carriers and waveguide have boundaries on both sides of the active region. In the separate confinement heterostructure (SCH), the carriers are confined in a region within the waveguide.

A cross section of the T912 double heterojunction injection laser diode is illustrated in figure 9–5. Note that the active area is aluminum gallium arsenide (AlGaAs), while one boundary to the active region is N-type AlGaAs and the opposite boundary to the active region is P-type AlGaAs. On the top of the structure is a layer of P-type gallium arsenide (GaAs), while on the base is an N-type gallium arsenide (GaAs) substrate. The T912 device is of a stripe geometrical design; that is, a layer of silicon dioxide (SiO_2) is placed between the junctions and a copper sink. The purpose

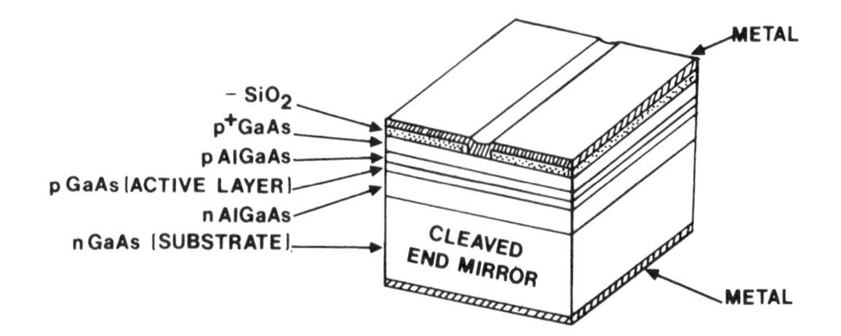

Figure 9–5 Heterostructure PN-Junction Laser

of the SiO_2 is to provide a narrow stripe rather than a wide structure that may eliminate noise.

The output characteristics of the T910 are compatible with PIN and APD detectors and photocathodes. They are ideally suited for use as light sources in high-performance, long-distance, high-bandwidth fiber-optic communication systems. They may also be used as target designators, intrusion alarms, and other applications.

ILDs used for fiber optics operate in the 810- to 850-nm wavelength range. Typical of the ILDs is the ITT series T910–T912 series shown in figure 9–3. Some of the series specifications are as follows:

Peak Emission Wavelength	840 nm
Spectral Width	4 nm
Peak Forward Voltage at 190 mA Forward Current	2 V
Peak Reverse Voltage	2.5 V
Continuous Forward Current	200 mA
Pulsed Forward Current	200 mA
Operating Temperature Range	0 to 75°C
Optical Power Varies from 0.5 to 5 mW Depending on the Model	
Emitter Diameter	1×20 μm

THE DETECTOR

The light detector accepts a light signal from the optical fiber and converts it into an electrical current. There are two types of detectors in use: the PIN diode and the APD. The PIN diode (contains positive, intrinsic, and negative solid-state layers in its construction) exhibits fast rise time and acceptable bandwidth parameters. It is reasonably priced. The APD diode (avalanche photodiode) exhibits fast rise time and acceptable bandwidth parameters. The APD is more expensive than the PIN diode since it provides greater receiver sensitivity. The APD also requires an auxiliary power supply.

A detector material is specially selected so that when it is exposed to light rays it will absorb the light energy. If the detector responds to the light energy (flux), and not the wavelength, the detector is said to be nonselective. If the detector responds by varying detection at different wavelengths, it is selective. Selectivity or responsivity is the detector's response per unit of light. Wavelength responsivity is called *spectral response*. Frequency response is the speed by which the detector responds to changes in radiation amplitude. Fluctuations in output current and/or voltage are referred to as *noise*. Noise is usually caused by current that flows in the detector regardless of whether light is applied. Current, such as this, is termed *dark current*, because it flows even without radiation. A common specification for a detector is the signal-to-noise ratio. This ratio is the noise current divided by the signal current.

THE PIN DIODE

In fiber-optic technology, signal processing requires some modification. It is the purpose of the PIN diode to serve as a transducer to convert the detected light (optical power) to electrical current. Conventional photodiodes respond too slowly to light signals. This problem has been cured by the PIN diode. The PIN diode is so called because of the layer material by which it is constructed. The word PIN is an acronym for P-type, intrinsic, N-type materials. A PIN photodiode is one in which a heavily doped P region and a heavily doped N region are separated by a lightly doped I region. In figure 9–6 the center of the structure represents the I region. The

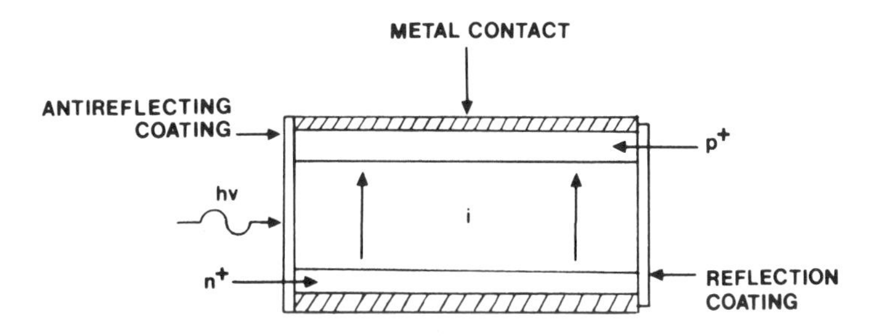

Figure 9–6 Structure of a PIN Diode

resistance of the I region can range from 10 to 100,000 ohms per centimeter. The P and N regions are less than 1 ohm per centimeter. Since a depletion area can extend farther into a nondoped or lightly doped region the PIN photodiode has an extremely large depletion area. This large depletion area provides the PIN photodiode with much faster speeds, lower noise, and greater efficiency at longer wavelengths.

Typical PIN diodes have the visual appearance of photodiodes. The shapes are modified for the application just as other source or detector components. Figure 9–7 is typical of the photodetectors used today.

The typical specifications which surround the PIN diode are as follows:

Responsivity: 0.5 A/W

Responsivity is the ability of the detector to convert light power to electrical current. This is expressed in amperes per watt.

Bandwidth: 840 nm

Bandwidth is matched to the light source in the application. A good detector should have very high response and sensitivity at the wavelength of the source producing the light signal. Speed of response should be sufficient to accommodate the rate of information applied.

Figure 9–7 Typical Shape of a Photodetector

Noise is ultimately a measurement of the performance of a fiber-optic system. It degrades the signal and, in turn, the performance of the system. Quantum noise is inherent in the photodetection process. Photons must be dislodged and electron-hole pairing must take place. Dark current noise and leakage noise are due to diode imperfections that create unwanted displacement current. Beat noise is caused by modulation. It is small enough to be ignored. Thermal noise is caused by external components such as bias resistors and feedback resistors.

THE AVALANCHE PHOTODIODE (APD)

The avalanche photodiode (APD) is an improvement on the PIN diode for detection. The APD was devised to increase the detection sensitivity of the PIN. It does this by multiplying the number of electron pairs generated by incoming photons. The APD is a detector with the added attraction of internal amplification of the photocurrent.

The APD is constructed as illustrated in the figure 9–8. This particular APD is a side-illuminated structure. As you can see, the silicon material is P+, N, N+. It has a wavelength range of 0.4 to 1.1 μm. Germanium APDs may extend this wavelength to about 1.5 or 1.6 μm.

Operation of the APD is as follows. At low-reverse-bias conditions, the diode operates as a regular photodiode. As reverse bias is increased, carrier multiplication takes place. Carriers gain energy by moving through the high-field specially doped region. Carriers gain sufficient energy to create new electron-hole pairs through the process of impact ionization. The amount of multiplication of electron-hole pairs may be somewhat controlled by varying the reverse bias across the APD.

Specifications for the APD are similar to the PIN detector, as are its physical shape. APDs use considerably high bias voltage (100 V+). Responsivity for the APD is high, between 1 and 100 A/W. This specification

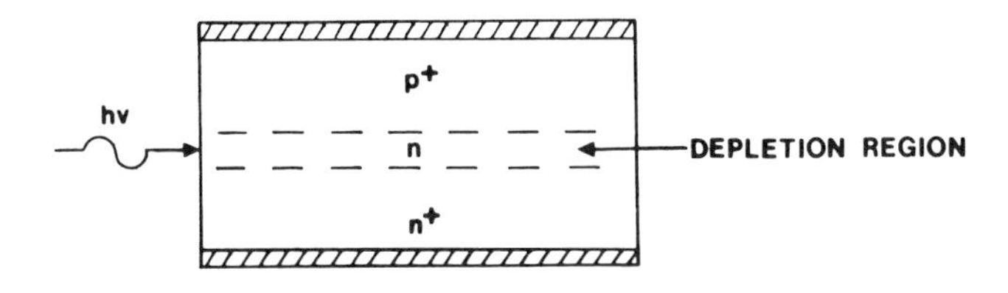

Figure 9–8 Structure of an APD Diode

is expressed in amperes per watt. Bandwidth is matched to the light source in the application.

The cost of the APD is considerable compared to the PIN diode. This and the fact that the APD requires gain stabilization against temperature variations may make the APD less desirable than the PIN diode. However, the APDs sensitivity and large responsivity may be overwhelming factors in the choice of a detector.

10

Philosophy of Fiber-Optic System Design

A simple fiber-optic system is called a *transmission link*. It consists of a transmitter with a light source, a length of fiber, and a receiver with a light detector. The basic operation of a system is to connect a digital or analog signal to a transmitter. Within the transmitter, the input signals are converted from electrical to optical energy by modulating an optical light source, normally achieved by varying the drive current. The modulated light is launched into a length of fiber, where it reflects from wall to wall through the fiber core. At the opposite end of the fiber a detector accepts the light and converts it back to an electrical signal. The electrical signal is converted back to its original form in the receiver. A brief discussion of each major component is provided.

TRANSMITTER

The purpose of the transmitter (driver) is to change the electrical signal into the required current to drive a low-impedance light source. The electrical inputs are either digital or analog. The choice of the converter should depend on the current requirement of the light source. If the signal is digital, the transmitter (driver) should consist of a high-speed pulser to turn the light source on and off. If the signal is analog, the transmitter (driver) should be able to supply current to the light source to transmit the positive and negative alternations of the signal.

RECEIVER

The function of the receiver is to accept low-level power from the detector and convert it into a high-voltage output. There are at least two methods of accomplishing this. The detector current produces a voltage drop across a load resistance. The voltage drop is directed into an amplifier. An output voltage representative of the transmitted signal is the result. The operational amplifier output voltage is the effect of the amplifier driving detector current through the feedback resistance. Again the output voltage is representative of the transmitted signal.

Other electronics may be added to the circuitry to maintain correct response. A gain control may be used on the front end to vary the impedance of the receiver. The operational amplifier is used as a current-to-voltage converter.

Since signal inputs are generally weak, shielding and power-supply decoupling are a requirement to achieve sensitivity. Sensitivity is set by input noise.

SYSTEM COMPATIBILITY AND SPECIFICATIONS

It stands to reason that if you acquire all the parts for a fiber-optic system from one company, the parts you buy will be compatible. Indeed, it may be intelligent thinking to do just that. Most manufacturers have developed complete systems and have off-the-shelf components available for the asking. Let's consider some of the things that are ultimately important when parts are chosen and a system is being created.

System Considerations

1. Digital
   a. Required bit error rate (BER) in bits per second. Upper bit error rate is usually in megabits per second (Mbps). Lower bit error rate is usually in bits per second (bps).
   b. Temperature operating range in degrees Celsius (°C)
2. Analog
   a. Bandwidth in hertz (Hz) or megahertz (MHz)
   b. Distortion in decibels (dB)
   c. Temperature operating range in degrees Celsius (°C)
3. Audio
   a. Bandwidth in hertz (Hz) or megahertz (MHz)
   b. Distortion in decibels (dB)
   c. Crosstalk in decibels (dB) (for multiple channels)
   d. Temperature operating range in degrees Celsius (°C)

4. Video
   a. Bandwidth in hertz (Hz) or megahertz (MHz)
   b. Distortion in decibels (dB)
   c. Crosstalk in decibels (dB) (for multiple channels)
   d. Temperature operating range in degrees Celsius (°C)

Transmitter Specifications

1. Input impedance in ohms (Ω)
2. Maximum input signal in volts dc (V_{dc}), volts effective (V_{rms} or V_{eff}), volts peak to peak ($V_{p\text{-}p}$)
3. Optical wavelength in micrometers (μm) or nanometers (nm)
4. Optical output power in microwatts (μW)
5. Optical output rise time in nanoseconds (ns)
6. Required power supply in volts dc, usually $5 \pm 0.25\ V_{dc}$ or $\pm 15 \pm 1\ V_{dc}$

Light-Source Specifications

1. Continuous forward current in milliamps (mA)
2. Pulsed forward current in milliamps (mA)
3. Peak emission wavelength in nanometers (nm)
4. Spectral width in nanometers (nm)
5. Peak forward voltage in volts dc (V_{dc})
6. Reverse voltage in volts dc (V_{dc})
7. Temperature range in degrees Celsius (°C)
8. Total optical power output in microwatts (μW)
9. Rise/fall times in nanoseconds (ns)

Fiber Specifications

1. Mode—single or multimode
2. Index—step or graded
3. Attenuation in decibels per kilometer (dB/km)
4. Numerical aperture (a sine value)
5. Intermodal dispersion in nanoseconds per kilometer (ns/km)
6. Core index of refraction (a ratio)
7. Cladding index of refraction (a ratio)
8. Core diameter in micrometers (μm)
9. Cladding diameter in micrometers (μm)
10. Tensile strength in pounds per square inch (psi)
11. Bend radius in centimeters (cm)

Cable Specifications

1. Numbers of fibers (a unit)
2. Core diameter in micrometers (μm)
3. Cladding diameter in micrometers (μm)
4. Cable diameter in millimeters (mm)
5. Weight in kilograms per kilometer (kg/km)
6. Minimum bend radius in centimeters (cm)

Detector Specifications

1. Continuous forward current in milliamps (mA)
2. Pulsed forward current in milliamps (mA)
3. Peak reverse voltage in volts dc (V_{dc})
4. Temperature range in degrees Celsius (°C)
5. Optical power output in microwatts (μW)
6. Threshold current in milliamps (mA)
7. Rise/fall times in nanoseconds (ns)
8. Radiation pattern in angular degrees

Receiver Specifications

1. Output impedance in ohms (Ω)
2. Output signal level in volts dc (V_{dc}), volts effective (V_{rms} or V_{eff}), volts peak to peak ($V_{p\text{-}p}$)
3. Optical sensitivity in microwatts (μW), nanowatts (nW), decibels (dB), or megabits per second (Mbps)
4. Optical dynamic range in decibels (dB)
5. Analog output overload in percent (%)
6. Analog output rise time in nanoseconds (ns)
7. Digital output rise time in nanoseconds (ns)
8. Required power supply in volts dc, usually $5 \pm 0.25\ V_{dc}$ or $\pm 15 \pm 1\ V_{dc}$

DESIGN CONSIDERATIONS

Before any thoughts of developing an optical fiber system can be completed, certain factors must be realized. First, the signal information must be known. Is the signal analog or digital? What is the information bandwidth? What power is required? Second, length of the transmission line has to be determined. How far is it between transmitter and receiver? Are there any physical obstacles that must be skirted or must the cable go through anything? Finally, what are the tolerable signal parameters? What is the acceptable signal-to-noise ratio (SNR) if the system is analog? What is the acceptable bit error rate (BER) and rise/fall time if a digital system? Once the basic parameters of the system are set, the system development can take place.

Design Procedures

Procedures for design of an optical fiber system are as follows:

1. Determine the signal bandwidth.
2. Determine the signal-to-noise ratio (SNR) if the signal is analog. This is the ratio of output signal voltage to noise voltage. The ratio is expressed as 10:1, 8:1, 20:1, and so on. The largest signal-to-noise ratio is desirable. The SNR is expressed in decibels (dB). SNR curves are provided on detector data sheets.
3. Determine the tolerable bit error rate (BER) if the signal is digital. BER is the ratio of incorrect bits to total bits of data. A typical good BER is 10^{-8}. BER curves are supplied by the manufacturers of detectors.
4. Determine link distance, that is, the distance between the transmitter and the receiver.
5. Select a fiber based on attenuation.
6. Calculate fiber bandwidth for the system. This is accomplished by dividing the bandwidth factor in megahertz per kilometer by the link distance. The bandwidth factor is provided on fiber manufacturers' data sheets.
7. Determine power margin. This is the difference between light source output power and receiver sensitivity.
8. Determine total fiber loss by multiplying fiber loss in decibels per kilometer by the length of the link in kilometers.
9. Identify the number of connectors. Multiply the connector loss (provided by the manufacturer) by the number of connectors.
10. Identify the number of splices. Multiply the splice loss (provided by the manufacturer) by the number of optics.
11. Allow 1 dB for detector coupling loss.
12. Allow 3 dB for temperature degradation.
13. Allow 3 dB for time degradation.
14. Sum the fiber loss, connector loss, splice loss, detector coupling loss, temperature degradation loss, and time degradation loss (add values of steps 8 through 13) to find total system attenuation.

15. Subtract total system attenuation from power margin. If the difference is negative, the light-source power receiver sensitivity must be changed to create a larger power margin. A fiber with a lower fiber loss may be chosen, or the use of fewer connectors and splices may be an alternative, if it is possible to do so without degrading the system.
16. Determine rise time. To find the total rise time, add the rise time of all critical components, such as the light source, intermodal dispersion, intramodal dispersion, and detector. Square the rise times. Then take the sum of the total squares. Square this sum and multiply it by some parameter factor such as 110% or 1.1, as in the following example.

$$\text{System rise time} = 1.1\sqrt{T_1^2 + T_2^2 + T_3^2 + \cdots}$$

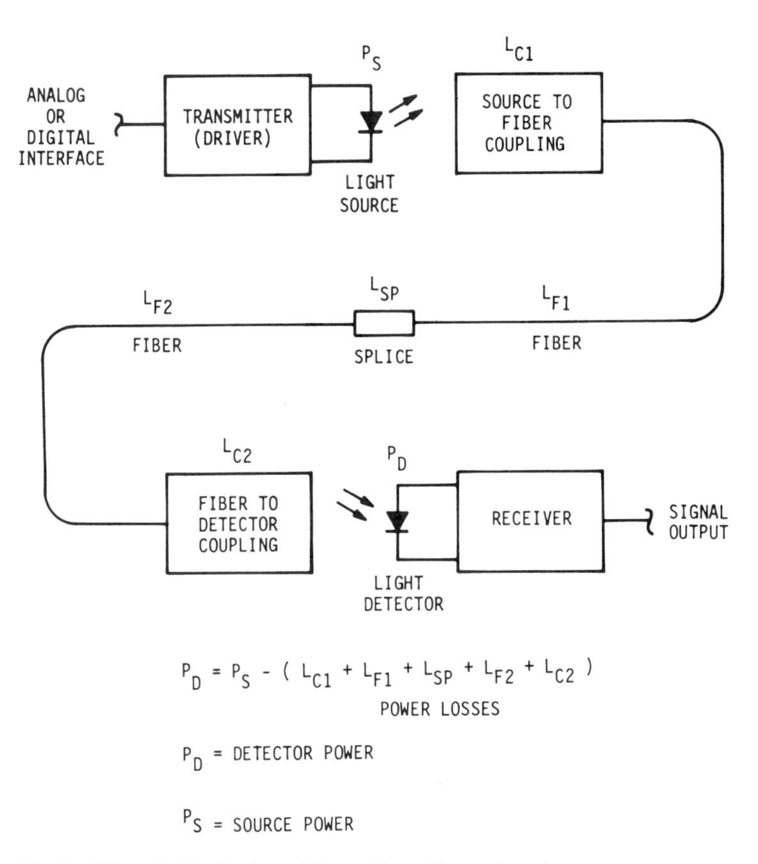

Figure 10–1 Fiber-Optic System Attenuation (Power Loss)

Fiber-Optic System Attenuation

The total attenuation of an optical fiber system is the difference between the power leaving the light source and the power entering the detector. In the figure 10–1, power entering the fiber is designated as P_S or source power. L_{C1} is power loss at the source to fiber coupling, usually 1 dB per coupling. The power is of that signal launched into the fiber from the light source at the fiber coupling. L_{F1} represents the loss in the fiber between the source and the splice. Fiber losses are listed in specifications and have power losses at about 10 dB/km. L_{SP} represents the power loss at the splice. A representative power loss of a splice is 0.3 to 0.5 dB. L_{F2} represents power loss in the second length of fiber. L_{C2} is the power loss at the fiber-to-detector coupling. Finally, P_D is the power transmitted into the de-

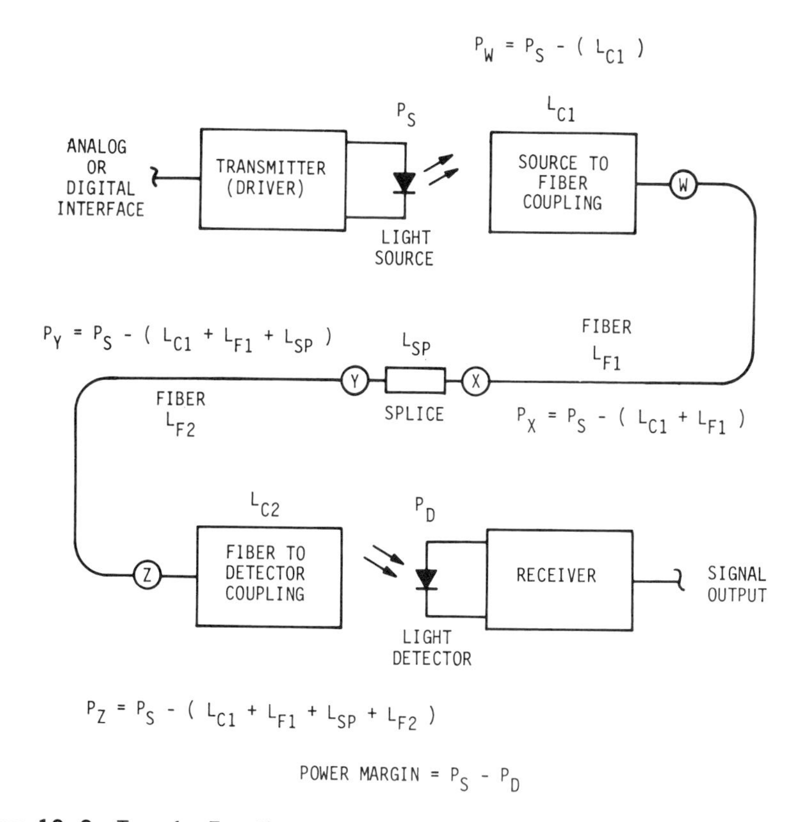

Figure 10–2 Transfer Functions

tector. Other power losses due to temperature and time degradation are generally around 3-dB loss each. Power at the detector is then generalized as

$$P_D = P_S - (L_{C1} + L_{F1} + L_{SP} + L_{F2} + L_{C2})$$

Note: All power and losses must be expressed in decibels (dB).

Transfer Functions

Transfer functions are the power formulas that describe the power at specific points along the transmission line. In figure 10–2, P_S is representative of the power output from the light source. Power at different points can be represented by a power formula called a *transfer function*. Some typical transfer functions applicable to the fiber-optic system in figure 10–2 are listed in table 10–1.

TABLE 10–1 Table of Transfer Functions (See Figure 10–2)

Parameter	*Transfer Function*
Power out of light source	P_S
Power into fiber at point W	$P_W = P_S - L_{C1}$
Power into splice at point X	$P_X = P_S - (L_{C1} + L_{F1})$
Power into fiber at point Y	$P_Y = P_S - (L_{C1} + L_{F1} + L_{SP})$
Power into fiber at point Z	$P_Z = P_S - (L_{C1} + L_{F1} + L_{SP} + L_{F2})$
Power into detector	$P_D = P_S - (L_{C1} + L_{F1} + L_{SP} + L_{F2} + L_{C2})$

11

Communication Links

The unique advantages of the fiber-optic systems are in greater bandwidths, small cable cross sections, and isolation from electromagnetic radiation. Probably the most significant of all the applications of fiber-optic systems are in the telephone industry. Substitutions for electrical conductors are being used throughout the country and, indeed, the world. Standardization within the industry is made possible by the North American Digital Hierachy, which is the standard in the United States. Essentially, the standards are in channels and bit rates. These are listed in table 11–1. Applications in telephone are intercity trunks, interoffice trunks, and local loops. The key advantages for telephone systems are high bandwidth, low attenuation, and small size.

TABLE 11–1 North American Digital Hierarchy Standards

System Line	*4-kHz Voice Channels*	*Bit Rate (Mbps)*
T1	24	1.544
T2	96	6.312
T3	672	44.736
T4	4032	274.176

Broadband networks such as television are using fiber waveguides in commercial cable television (CATV), wired city (community), and dedicated communication systems. The advantages are simpler transmission-line characteristics and noninterference (security).

Computers for general electronic data processing and computer-based process control and instrumentation systems represent large-scale use of optical waveguides. At a given bit rate, the benefits of optical waveguides versus metallic conductors increase with distance. This, together with optical waveguides' immunity to electromagnetic interference, suggests the greatest motivation to apply waveguides in computer systems operating in a high-electromagnetic-interference (EMI) environment. Computer applications for fiber optics are dispersed systems, intersystem wiring, computer-based industrial process control, and instrumentation. The advantage is immunity to interference.

Military applications include wiring of weapons systems such as aircraft, helicopter, ship, or submarine. Technically, many of them entail data transmission within computer-based systems. Headquarters, field, and base communications are of special importance to the Army. Surveillance systems operated from shore, ship, or submarine may make a substantial impact on the Navy's capabilities to detect, identify, and evaluate underwater threats. Secure communications like those for missiles are receiving attention. Having recognized the potential of optical waveguides very early, the military laboratories and their contractors have an impressive number of programs under way. These programs are yielding significant technical results and are expected to conclude with a line of military-qualified hardware, demonstrated system feasibility, and know-how to apply the technology of optical waveguides. Again, the advantages are immunity to electromagnetic interference and electromagnetic pulses (EMP), security, and the size/weight factors.

LARGE FIBER-OPTIC INSTALLATION

One of the major installations using fiber optics today is the Lake Placid Winter Olympics Lightwave System. The system was a joint effort of New York Telephone, AT&T, Western Electric, and Bell Labs. (See figure 11–1.) The new installation was obviously a tremendous success considering the clarity and quality of the television broadcasts of Winter Olympics 1980. Problems were few and primarily mechanical. The new installation was in place and operational by October 1979. In the figure the system connection is shown. Its first purpose was to transform the Lake Placid tele-

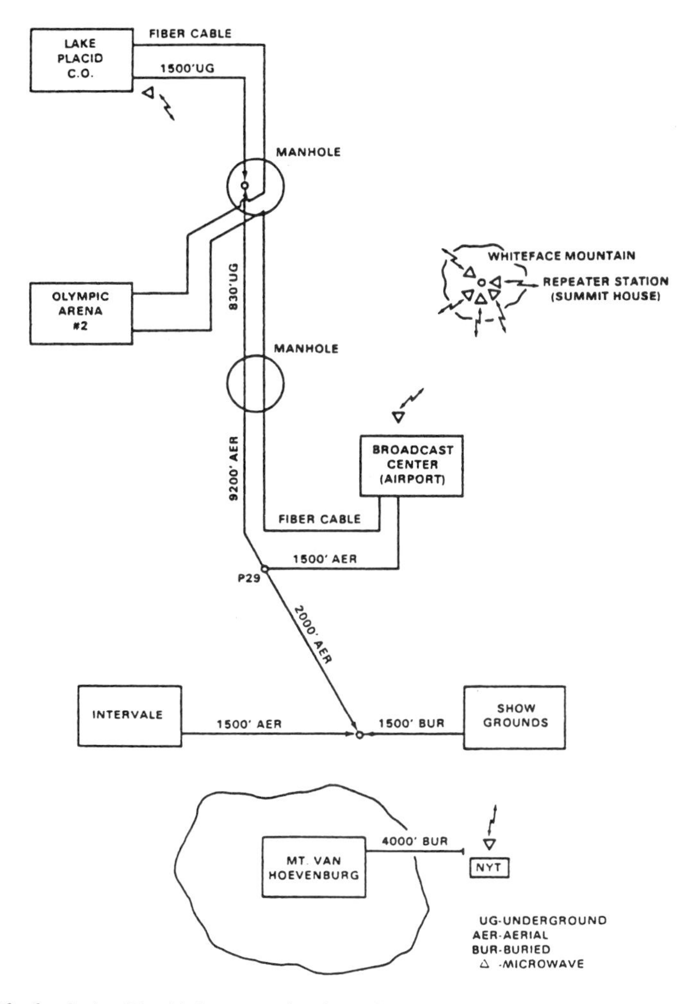

Figure 11–1 Lake Placid Communications System *(Courtesy New York Telephone)*

phone facility into an ultrasophisticated communications center capable of handling a wide range of telecommunications services necessary to support the 1980 Olympic events.

The lightwave system extends 2½ miles and links the Lake Placid telephone switching office, the Olympic ice arena, and the broadcast center. The broadcast center serves 25 mass-media agencies. The fiber cable, made by Western Electric, travels approximately 2 miles on poles (aerial) and the other ½ mile under ground. Part of the underground cable is buried. The cable consists of 12 glass fibers carrying digital voice signals and

Figure 11–2 The New York City and Westchester Fiber-Optic Telephone Link *(Courtesy New York Telephone)*

television. Six of the fibers carry 288 two-way voice conversations, while two of the fibers transmit the video and associated voice signals. The remaining four fibers serve as backup in the event of failure. An added part of the Olympic system is a microwave installation which serves the entire facility.

A second major effort using fiber optics is the permanent installation linking telephone company central offices between White Plains and East 38th Street, Manhattan, New York. (See figure 11–2.) This is an extremely long system (30 miles). It was installed to meet with a growing volume of telephone calls between New York City and Westchester. The New York system will carry both voice and data communications between company central offices at White Plains, Scarsdale, Tuckahoe, and Mount Vernon. This part of the major link will cover 11 miles and have a potential for handling 14,000 calls simultaneously. This part of the major effort was completed in 1980.

The second part of this major fiber-optic system will enter New York City from Mount Vernon, pass through city central offices, and terminate at the East 38th Street exchange. This second link will cover over 17 miles and is installed in existing cable ducts. The Mount Vernon–mid-Manhattan fiber-optic link is probably the beginning of major usage of fiber optics in the telephone industry.

A BASIC T1 FIBER-OPTIC SYSTEM

Early in 1979, Southern Bell Telephone and Telegraph and Florida Power and Light completed an installation of a T1 system. You may recall that a T1 system consists of 24 voice-frequency channels into a 1.544 Mbps pulse code modulation (PCM) bipolar digital-bit stream. The system was manufactured by ITT Telecommunications of Raleigh, North Carolina. The unique feature of the system is the use of optical cable to metallic cable conversion in a repeater manhole environment. (See figure 11–3.) The con-

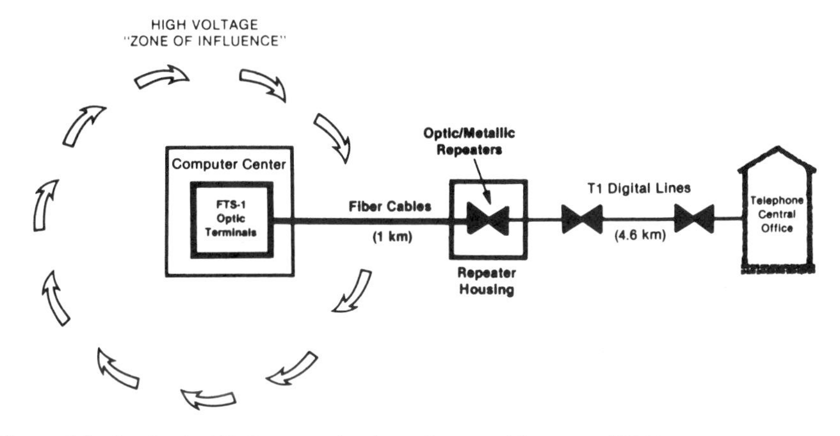

Figure 11–3 Basic T1 Communications System *(Courtesy ITT Telecommunications)*

version repeaters convert infrared light pulses into electronic pulses for transmission over conventional digital span lines to the central telephone office.

System Description

The computer center consists essentially of three pulse-code modulation (PCM) channel banks, an automatic protection switch (APS) for three service lines with a spare, an optics shelf consisting of optical transmitter units (OTUs), office power converters (OPCs), and optical receiver units (ORUs). (See figure 11–4.)

Each PCM channel bank multiplexes 24 voice-frequency (VF) channels into a 1.544-Mbps pulse-code modulation bipolar digital-bit stream for transmission. A special test-and-alignment unit plugs into the shelf of the PCM units to measure and adjust the receiver voice-frequency (VF) levels, provide a 1020-Hz test tone for setting transmit voice-frequency levels, and to measure idle channel noise, linearity, distortion, crosstalk, hybrid balance, and office equipment losses.

The automatic protection switch accepts the 1.544-Mbps signals from the three PCMs. If any of the three optic service lines fails, the APS automatically transfers PCM signals to the spare line. After the fault is cleared, reset to normal is automatic. The optic line shelf consists of optical transmitter units, office power converters, and optical receiver units. The optical transmitter units convert the 1.544-Mbps second-rate bipolar

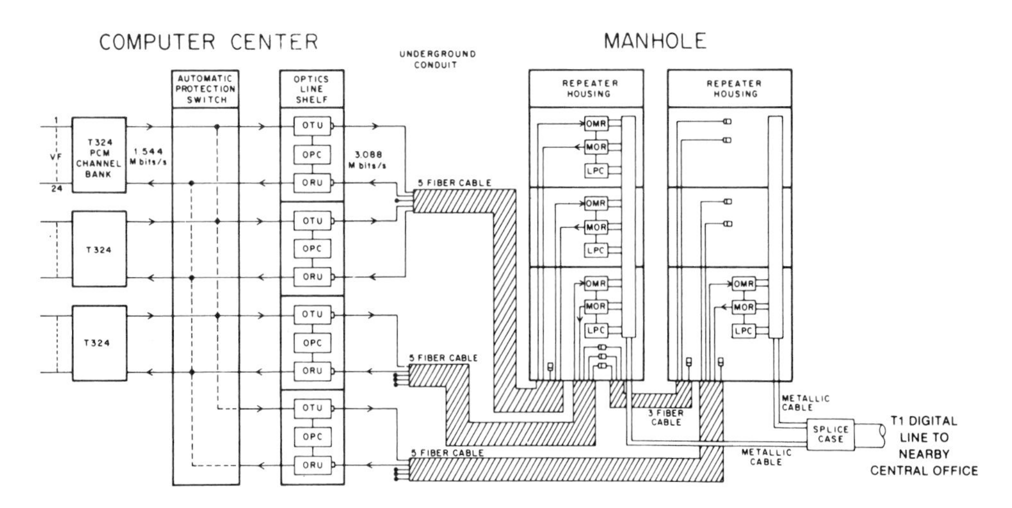

Figure 11–4 T1 System Description *(Courtesy ITT Telecommunications)*

electrical signal to a 3.088-Mbps optical signal suitable for launching over the fiber-optic line. Wavelength of the launched signal is 850 nm. The optical receiver unit detects the 3.088-Mbps signal and decodes it back into a 1.544-Mbps bipolar signal. The office power converter provides the power for operation of the OTU and the ORU.

Three optical cables with five fibers each are routed through underground conduit from the optical line shelf to repeater housings in the manhole. The cables are 7 mm in diameter. The fibers are graded index and have an attenuation of 6 dB/km.

An optical-to-metallic conversion repeater interfaces the opposite end of the fiber cables to the T1 span line. Each converter has capability of handling three single system repeaters. In this system the two repeaters are also connected together with three fiber cables allowing for expansion. The converter housing is stainless steel and is pressurized to 15 psi to purge moisture. Each conversion repeater consists of three sets of conversion units. Each set has an optical-to-metallic regenerator (OMR) that consists of a 3.088-Mbps optical receiver, a bit-rate converter, and a 1.544-Mbps metallic span-line driver. The metallic-to-optic regenerator (MOR) contains a 1.544-Mbps metallic regenerator, a bit-rate converter, and a 3.088-Mbps optical transmitter. The line power converter (LPC) is a dc-to-dc converter which provides power for the OMR and MOR units. All the units in the conversion repeater are similar in function to the units on the optical line shelf. Metallic cables connect from the conversion repeaters to the T1 digital line in the central office.

A BASIC T3 SYSTEM

A typical T3 communications system spanning a distance of 22 km was scheduled for completion by the Pennsylvania Commonwealth Telephone Company in mid-1979. The system, manufactured by ITT Telecommunications, utilizes four intermediate repeaters between two north-central Pennsylvania communities of Wellsboro and Mansfield. The system carries toll, intertoll, operator, and special service traffic. Optical cables include 10 km of direct buried (plowed) cable, 1.0 km of underground (duct) cable, 3.0 km of aerial cable lashed to existing cable, and 8.0 km of aerial cable lashed to a new messenger strand.

System Description

The modified T3 communications system (see figure 11–5) accepts a 44.736-Mbps electrical signal from a digital radio or a multiplexer switch and converts it to an optical signal. The optical signal is launched into fiber-optic cables and transmitted to a receiver at the far-end optical terminal. Sufficient optical repeaters are placed into the link between the two terminals. The receiver detects the optical signal, converts it to an electronic level, and forwards it to the receive side of the far-end terminal, where it is converted to T3-level signals and routed to radio or multiplexer.

In the lower part of the figure 11–5, the metallic equivalent of a T1 communication system is compared to the optical system. Electrical signals from digital radio or a digital switch are fed to repeater and transmitted over metallic lines to repeaters at the far-end central office and routed to radio channels. A typical T3 system conversion then consists of a pair of office terminal repeaters and their housings and the fiber-optic cables with connections.

A T3 office terminal is pictured in figure 11–6. Internally, the terminal consists of an optical transmitter, an optical receiver, a looping unit, scrambler and descrambler, and an alarm system. See figure 11–7 for a line drawing of a functional block diagram of the T3 office terminal.

A T3-level signal is routed to and from the T3 terminal. The signal is scrambled to randomize the input, which improves repeater and receiver clock extraction and removes jitters. Parity bits are added to allow bit-

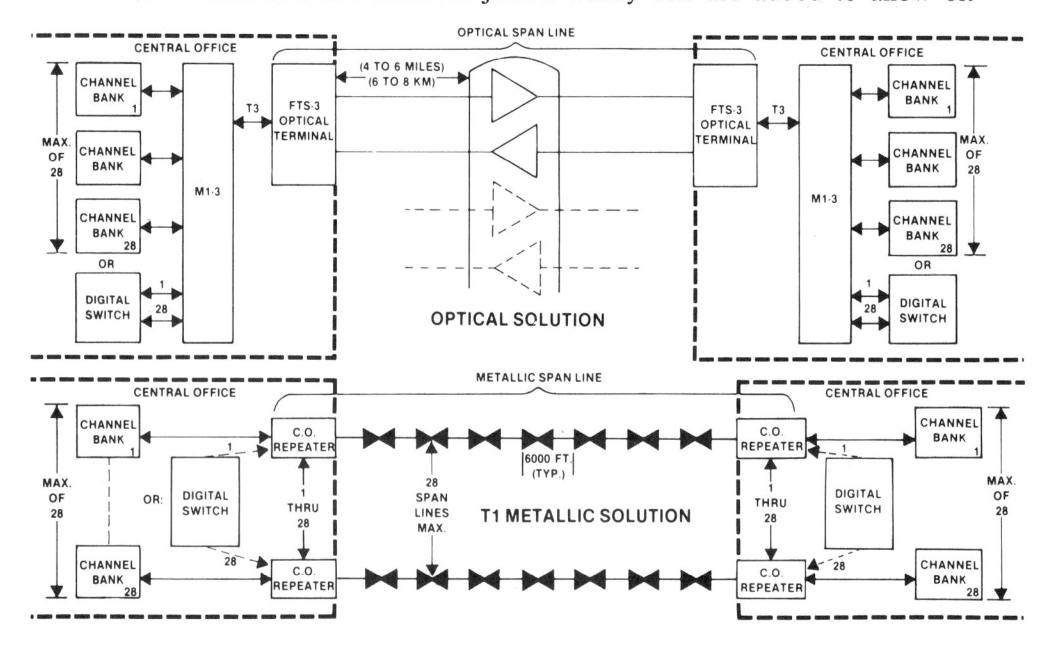

Figure 11–5 Basic T3 Communications Systems *(Courtesy ITT Telecommunications)*

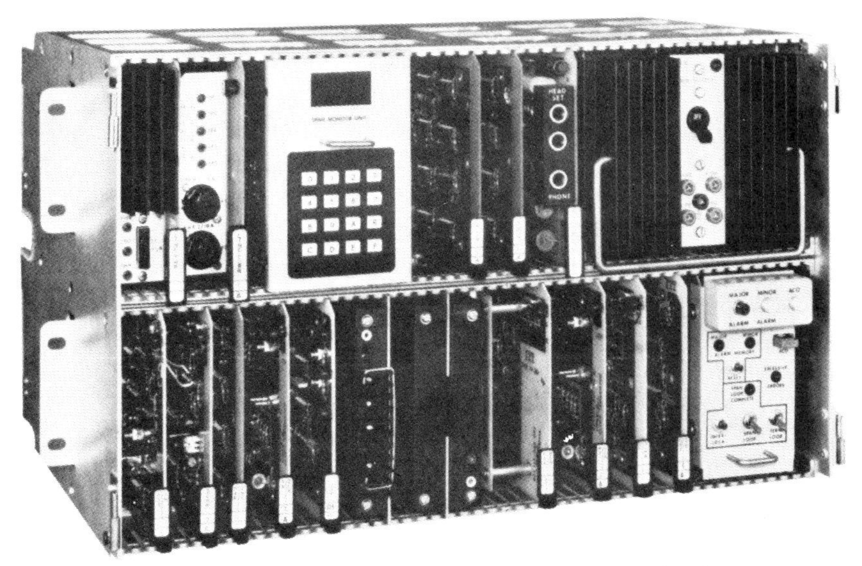

Figure 11–6 T3 Office Terminal *(Courtesy ITT Telecommunications)*

error-rate (BER) monitoring and the transmission of system control functions. The 44.736-Mbps signal (now a 47.367-Mbps signal) is routed through the looping unit and to the optical transmitter for transmission. The looping unit allows out-of-service diagnostic testing from either end of

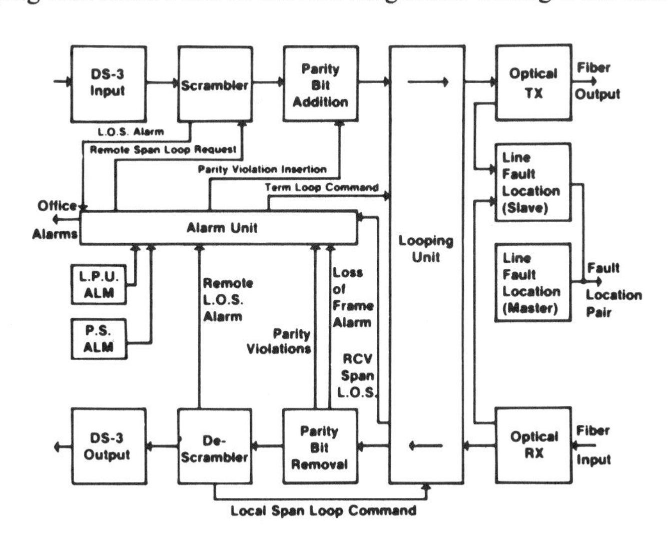

Figure 11–7 Functional Diagram of T3 Office Terminal *(Courtesy ITT Telecommunications)*

the transmission link. One operator at one end of the system can test all units from the input to the output except for the optical receiver and transmitter.

The optical transmitter modulates an injection laser diode (ILD) at a near-infrared wavelength of 850 nm. Modulated light is launched into fiber cables and transmitted toward the far-end receiver. The modulated light moves through the fiber cable (through repeaters) and to the optical receiver in the far-end office terminal. The optical receiver employs an avalanche photodiode (APD) as a detector. The signal is amplified. Clock-timing information is extracted and the signal data stream is regenerated. Parity bits added at the input of the opposite-end office terminal are removed and signals are descrambled. T3-level signals are then routed to radio or multiplexer. All signals along within the office terminal are monitored in the alarm unit and line-fault information is collected between the optical transmitter and receiver.

Cable Routing Layout

The illustration in figure 11–8 is a line drawing of the fiber-optic cable and its metallic counterpart. Note that the expanse between metallic splices is typically 2 km while the distance between optical cable splices is 1 km. The practical loss allocations, together with a reasonable safety margin, result in a 5-km expanse between the optics terminal shelf in the T3 office terminal and the repeater housing. Fiber cables are spliced in the field using a flame-fusion process. A fiber position fixture and a microscope are utilized to control alignment of fiber ends. Splice losses are around 0.5 to 1.0 dB. Considering that the splicing takes place in the field, in varying kinds of weather, the results are more than satisfactory.

Optical Repeater Housing

Repeater housings like the one in figure 11–9 are designed to be installed on poles or in manholes. When pole-mounted, a sun shield is used to elim-

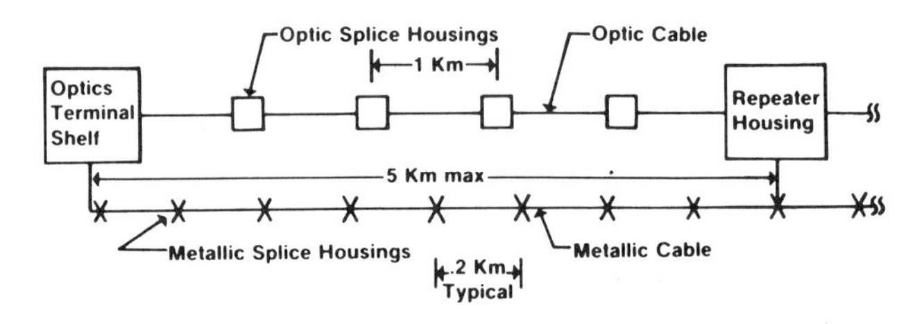

Figure 11–8 T3 System Cable Routing *(Courtesy ITT Telecommunications)*

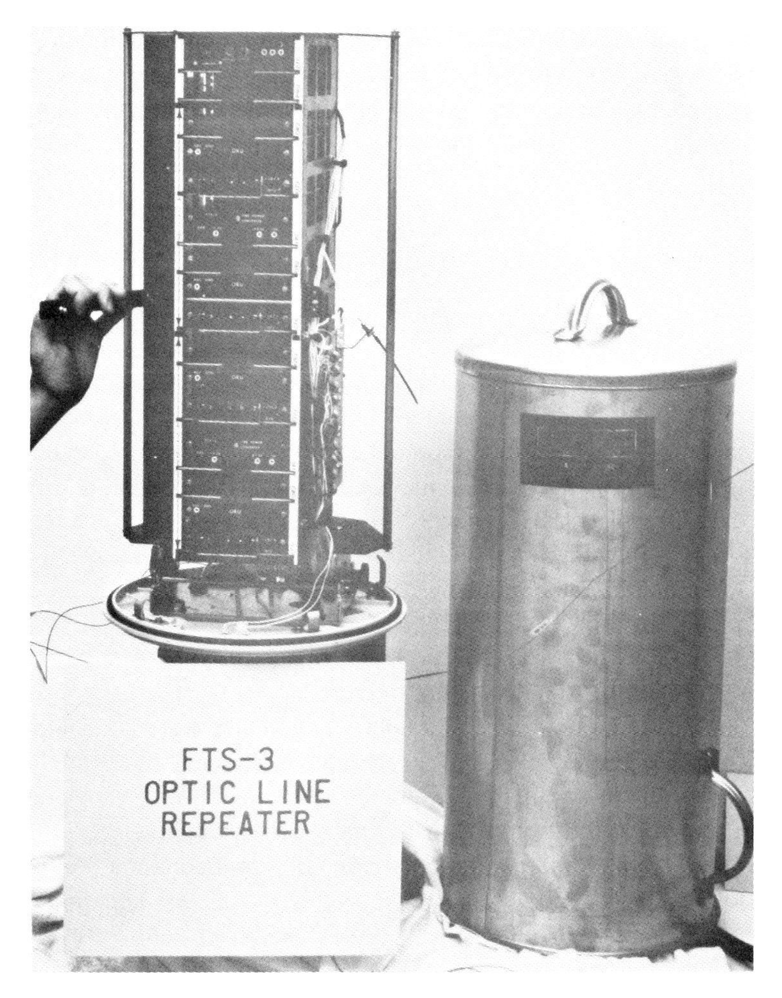

Figure 11–9 T3 System Repeater *(Courtesy ITT Telecommunications)*

inate heat rise that would reduce laser life. The housing contains the same optical receivers and transmitters as those in the office terminal. The case is pressurized. It has an internal electronic card cage designed to hold two bidirectional optical repeaters, power supplies, and repeater performance monitoring and reporting systems. The housing is equipped with metallic stubs for power leads, order wire, and system-monitoring wire pairs. Load coils and lightning protection for the copper pairs are also provided. Optical cables are installed in the field and with an epoxy dam. Power for the repeaters is ±130 V dc to provide a 140-mA loop current. Current is routed to the repeater using a 22- or 24-gage copper pair. At the present time, optical span line electronics cost about the same per voice-channel mile as the metallic versions. Longer length and systems with large numbers of channels tend to make the optical solution a better purchase because of the cable cost savings.

Data Buses

The major use for fiber-optic multiple-access couplers is in multiuser, multiservice fiber-optic buses. Within a typical bus are terminals, nodes, multiplexers, and links that comprise the network. In addition, the bus is separated into access and center user networks. Local users are grouped and interact in access networks. Center users are interconnecting. Services that can be handled on a fiber-optic bus include digital data, voice, video, and informational retrieval. These services can be segregated by frequency-division multiplexing. High-bandwidth fiber enables many channels to be multiplexed on a single-fiber bus.

Data-Bus Multiplexing

After the multiple-access couplers have been selected, a method for multiplexing several messages onto a single channel and a method to control and route data traffic are required. Multiplexing techniques in conventional wideband radio frequency (RF) networks are characterized by modulating carrier waveforms in the time domain. Two common methods of carrier modulation are time-division multiplexing (TDM) and frequency-division multiplexing (FDM).

TYPICAL SPECIALTY CABLE AND ITS APPLICATION

The type AT optical fiber cable is typical of the cables used throughout the communication industry. (See figure 11–10.) The AT cable was developed by General Cable Company and installed by General Telephone of California for use in what was reported to be the first use of optical fiber cable for regular telephone use. The cable was installed in a General Telephone route between the Artesia Central Office and the Long Beach Toll Center. Two regenerators (boosters) were placed in the line. The entire length of the system is 5.6 miles. One regenerator is 1.6 miles from the Long Beach Toll Center and the second regenerator 1.7 miles from the Artesia Central Office. The distance between the regenerators is 2.3 miles. A line drawing of this layout is shown in figure 11–10.

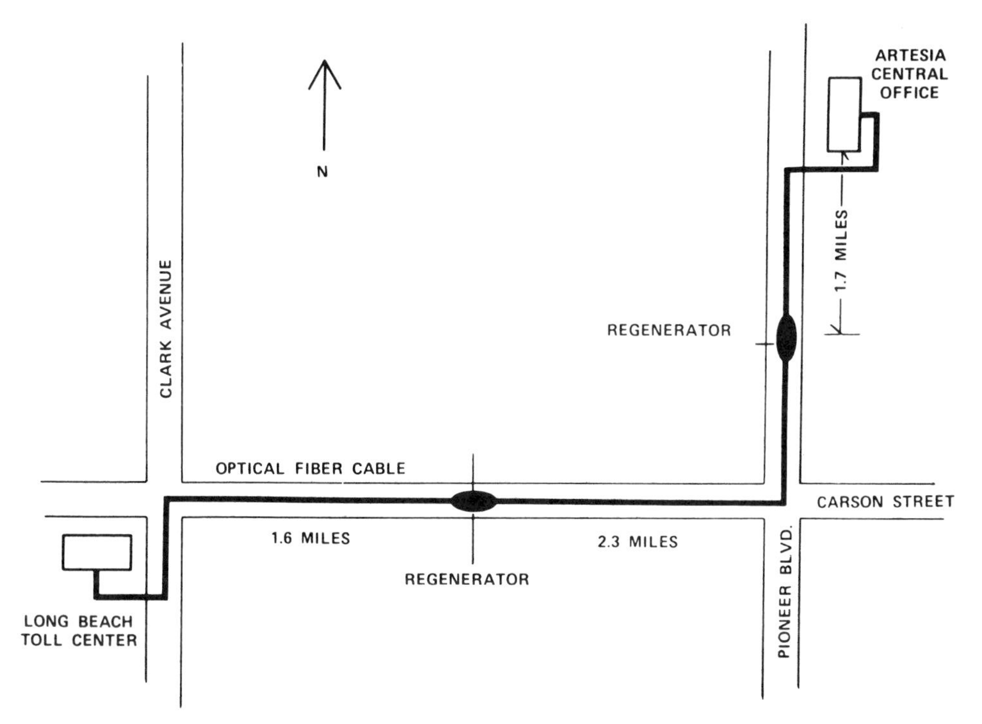

Figure 11–10 Typical Communications Fiber-Optic Cable Layout *(Courtesy GTE Corp.)*

AT Cable Construction

A line drawing of the AT cable and its parts is illustrated in figure 11–11. The optical fibers are laid parallel to each other and incorporated into a flat ribbon between plastic tapes that can be stripped back to expose the fibers when necessary. The ribbon is about 6 mm in width and contains up to 12 fibers. The ribbon construction reduces the effects of microbending and also provides additional strength as well as simplified and much safer handling of fibers during cable manufacture and in subsequent splicing operations.

The ribbon containing the fibers is placed in a groove fashioned helically along the length of an extruded plastic core. The core is strengthened with a central copper wire that also serves as an electrical conductor. The cable sheath consists of a 1.9-cm-diameter aluminum tube of 1.3-mm wall thickness constructed from a strip bent and longitudinally welded in a continuous process to form a loose fit over the cable core. Covering the alu-

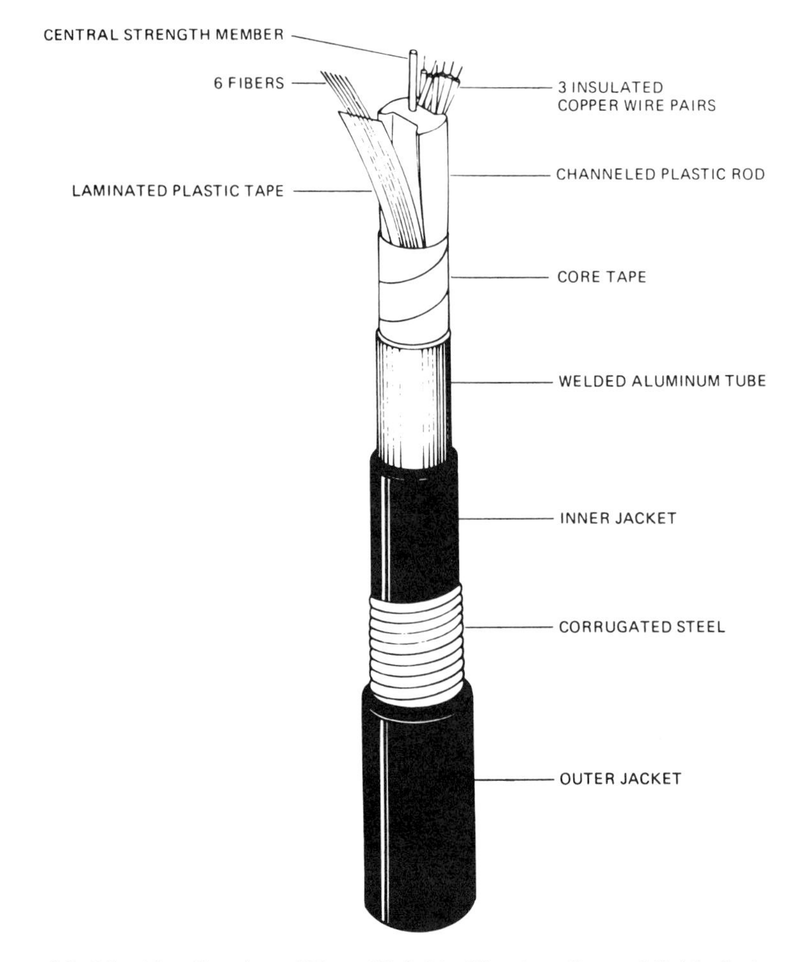

Figure 11–11 Line Drawing of Type AT Cable *(Courtesy General Cable Co.)*

minum tube is a polyethylene jacket, followed by a 0.15-mm corrugated steel tape with longitudinal overlip, and, finally, a second and outer polyethylene jacket. The metallic surfaces are flooded with an asphaltic anticorrosion compound. The production processes are continuous and place no restriction on the length of the cable that can be manufactured. The overall diameter of the cable is approximately 2.5 cm.

Other Cable Types

Figure 11–12 illustrates a more modern cable, the Gen Guide. This cable was developed by General Cable for a broadband communications link for

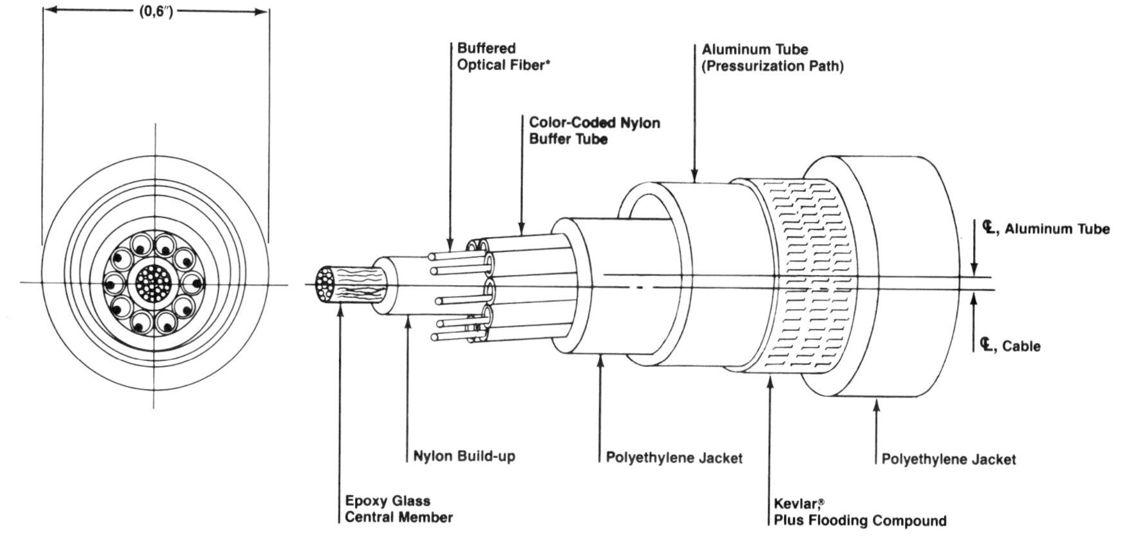

Figure 11–12 Type AT Cable *(Courtesy General Cable Co.)*

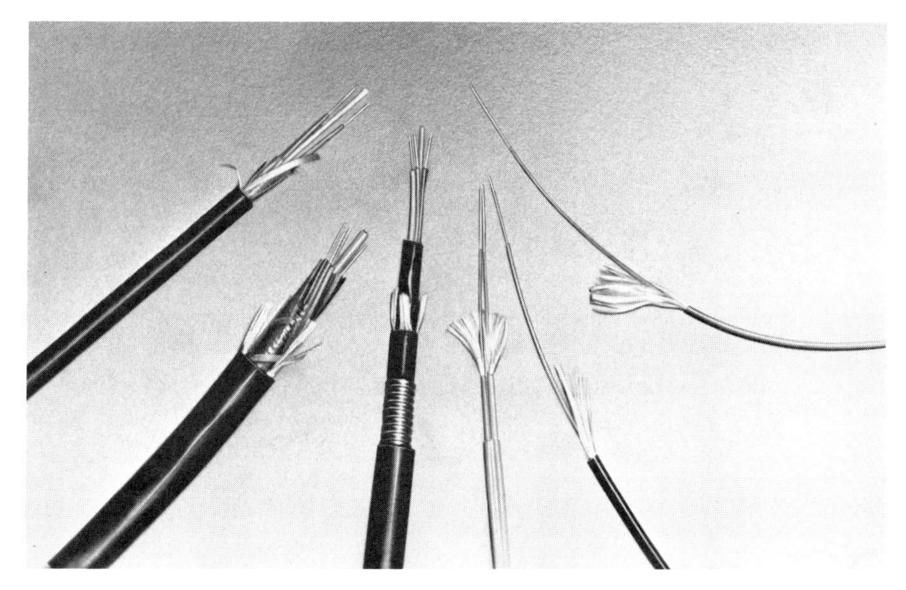

Figure 11–13 Typical Fiber-Optic Communication Cables *(Courtesy General Cable Co.)*

NASA at John F. Kennedy Space Center. Figure 11–13 illustrates six typical optical fibers used throughout the telephone and communications systems today. Each has special applications. From left to right in the photo, the following list describes some of their construction.

1. A non-metallic construction.
2. A cable with an aluminum shield under a jacket.
3. A cable with a corrugated steel armor between the inner and outer jacket.
4. A two fiber cable for indoors.
5. A single fiber patch cord for outside.
6. A single fiber patch cord for indoors.

12

Digital Data Processing Systems Interfacing

Fiber optics is no longer a curiosity. The use of fiber-optic cables to interface digital data processing equipment is now, and has been for some time, accepted practice. As with any technology, the user must carefully scrutinize the advantages and disadvantages of available alternatives. It is the purpose of this chapter to provide the reader with some of the basics of digital data fiber-optic systems interface. Further, typical applications will be supplied. To begin, we shall discuss the fundamental computer and some of its vernacular.

THE FUNDAMENTAL COMPUTER

The basic digital computer consists of three fundamental subsystems. (See figure 12–1.) These are the input and output (I/O), the central processing

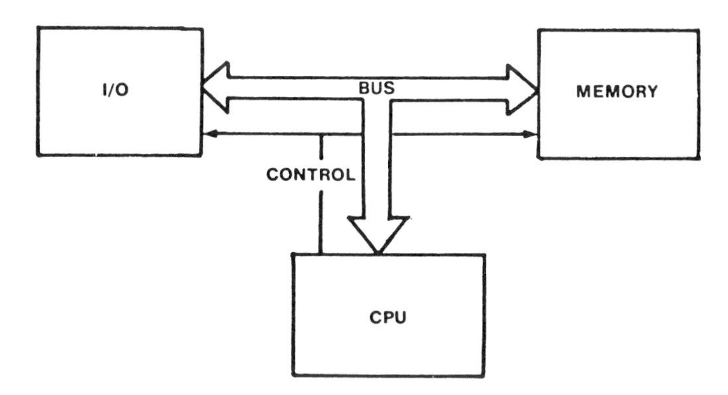

Figure 12–1 The Fundamental Computer

unit (CPU), and the memory. These are all tied together by electrical control wiring and signal data buses.

The I/O unit is an input/output device that allows an operator to interface with the computer. He/she may input data and/or programs and achieve results from a readout or a printer, etc. The memory unit stores the input, the program, and other data and gives it up when told to do so. The central processing unit performs mathematical and logical operations and controls the operation of all units. The data bus carries signal data between components. Control wiring carries commands to direct the system to function in its various modes.

SOME COMPUTER VERNACULAR

All computers have similar components and circuitry. Some vernacular can be applied to all computer systems.

Bandwidth The difference between the highest and lowest frequencies a circuit will encounter/pass/react to.

Baud A unit of data transmission speed equal to the number of code elements (BITs) per second.

BIT An acronym for BInary digiT. The smallest unit of information in the binary numbering system. The BIT is represented by the digits 0 and 1.

BIT Rate The rate at which binary digits, or pulses representing them, pass a given point in a communication line.

Bus A group of conductors considered as a single entity which connects various parts of a computer system.

Byte A sequence of binary digits usually operated upon as a unit.

Card Reader A device for reading information from punched cards.

Character One symbol of a set of elementary symbols, such as a letter of the alphabet or a decimal number.

Data A general term for any type of information.

Data Link Equipment, especially transmission cables and interface modules, which permits the transmission of information.

Dielectric An insulating material because of its poor electric conductivity.

External Memory Memory that is added to the basic computer and interconnected to the CPU is external memory.

Firmware There are often programs that are fixed within the computer. These programs are called read-only memory (ROM).

Hardware The physical parts of the computer are called hardware. These parts may be major components, such as the CPU, the memory, the I/O, and the keyboard. Other hardware may be the entire package or wiring. Connectors and printed circuits also fall into this category.

Modem Acronym for MOdulator/DEModulator. A device used to transmit and receive data by frequency shift keying (FSK). It converts FSK tones into their digital equivalent, and vice versa.

Module An interchangeable "plug-in" board containing electronic components which may be combined with other interchangeable items to form a complete unit.

Multiplexing The time-shared scanning of a number of data lines into a single channel. Only one data line is enabled at any instant.

Nonvolatile If memory is retained when power is turned off, the memory is termed nonvolatile.

Peripheral Equipment Units which may communicate with the basic computer but are not part of it (i.e., teletype, tape recorder, tape reader, etc.)

Printer A typewriter-like device used to print output information from a computer.

Programming Instructions input to the computer to carry out a sequence of operations are called programs. These are applied by keyboard, tape, punched cards, and other methods. Programming may be directed to the processing unit or the memory.

RAM Random access memory is that memory which may be read out of, written into, erased, and reused. In other words, it may be modified by almost any means.

ROM Read-only memory (ROM) is fixed programmed memory. It cannot be erased or modified. It may be called up (read out) but not written into. ROMs may be unplugged and replaced.

Software Instructions to the computer are called software. The instruction may be in the form of keyboard inputs to memory or processor, magnetic tape, punched cards, paper tape, and so on. Programmed instructions are loaded into the computer memory called random access memory (RAM).

Terminal A device similar to a typewriter which serves the operator of a computer as a tool to enter and extract information.

Volatile The word "volatile" has to do with loss of memory. If memory is lost when power is turned off, the memory is termed volatile.

SOME COMPUTER ACRONYMS

ANSI American National Standards Institute

ASA American Standards Association

AWG American Wire Gage

BCD Binary-Coded Decimal

BIT Binary Digit

COBOL Common Business-Oriented Language

CPU Central Processing Unit

CRT Cathode Ray Tube
CSA Canadian Standards Association
EDC Extended Data Comparison
EIA Electronics Industry Association
FDX Full Duplex (Simultaneous Transmit/Receive)
FORTRAN Formula Translation
FSK Frequency Shift Keying
HDX Half-Duplex (Two-Direction Transmit/Receive, but One at a Time)
IEEE Institute of Electrical and Electronic Engineers
I/O Input/Output
LED Light-Emitting Diode
LSD Least Significant Digit
LSI Large-Scale Integration
MCM Magnetic Core Memory
MSI Medium-Scale Integration
MODEM Modulator/Demodulator
MSD Most Significant Digit
MSI Medium-Scale Integration
OSHA Occupational Safety and Health Act
PC Programmable Controller
PCB Printed Circuit Board
RAM Random Access Memory
ROM Read-Only Memory
SSI Small-Scale Integration
TTY Teletype
TTL Transistor/Transistor Logic

LANGUAGE

There are many language types used with a computer. Each has its place in computer operation and understanding. Operation and programming are accomplished in computer language. The lowest-level computer language is machine language. Machine language is a string of 0's and 1's used to trigger the gates and multivibrators that control the computer operation. Machine language consists of two sets of instructions that instruct the operation to be used and the address to be operated on.

Assembly language is an instruction from machine language, translated into a mnemonic code such as CLR, which means clear. It is a memory aid and allows the programmer to construct easily remembered instructions.

COBOL is an acronym for COmmon Business-Oriented Language. FORTRAN is another acronym, meaning FORmula TRANslation. Obviously, this is a mathematically directed language. BASIC, along with FORTRAN and COBOL, are procedure-directed languages. That is, they are used to direct the computer to perform some logical task rather than to deal with machine operation. Such operations as retrieving data from memory or loading data are basic and typical functions of these higher-level languages.

THE OPTICAL FIBER INTERFACE

Current trends in the computer industry are leading toward process control and remote line entry. These trends demand increasing needs for data communication links. In the past, data entry and output and computation all took place in one room. Now, modern concepts have placed central processing units (CPUs) in one mainframe with satellite CPUs in remote places. The satellite CPUs have links to terminals, printers, card readers, and other specialized equipment.

Interconnecting of modern systems requires data links from 100 feet to 1000 miles or more. Short distances are owned or leased by the user. Long distances are usually provided by a telephone company.

Short-distance links use direct wire connection, line drivers, and limited-distance modems to transmit data. These are connected with twisted pairs or coaxial cables. These interconnecting cables must live in the hostile industrial environment. They are subjected to severe electromagnetic interference, such as large motors, high-voltage lines, and other industrial equipment. The EMI plus ground loops and communication crosstalk can and do cause intolerable data error rates.

Fiber-optic communication is immune to electromagnetic interference and generates no magnetic radiation, virtually eliminating crosstalk. Because the transmission medium is electrically nonconductive, ground loop and ground potential difference problems are eliminated, and fiber-optic cables can be routed adjacent to high-voltage lines.

Fiber-optic cables are much smaller and lighter than coaxial cables of equivalent capacity and offer a high degree of security against unauthorized monitoring. Also, they are unaffected by chemicals that are highly corrosive to metals, and cannot generate sparks.

Advantages of the Fiber-Optic Interface

Probably the greatest advantage of fiber optics is, as we have just discussed, its immunity to electromagnetic interference (EMI). This provides the user with a reliable product that has low crosstalk. This is also significant in that it reduces the bit error rate (BER).

Bit error rates in fiber-optic connections exceed 10^{-9}, compared to a 10^{-6} bit error rate for metallic connections. Bit error rate is defined as the ratio of incorrect bits to total bits in a received data stream. An increase in optical power will reduce the BER significantly. Obviously, the smaller the BER, the better the system performance.

Another major advantage of the optic fiber interconnect is its wide bandwidth potential; information carrying capacity increases with frequency. Laser transmitters may transmit data at speeds of 10^{14} bits per second. The lasers operate a monochromatic beam of intense light at frequencies 10,000 times that of the radio-frequency band, in the area of 1 to 2 gigahertz (1 to 2 billion cycles per second). This allows for high-speed transfer on the fiber-optic medium. It is now possible to multiplex 16-bit data buses onto a single fiber-optic cable, thus reducing connector hardware and wiring. This bandwidth capability far surpasses twisted pairs and coaxial cable methods.

Other advantages are space and weight savings, ease of installation, and rugged long life. A further advantage is that fiber-optic cables provide complete isolation between transmitters and receivers, which eliminates the need of a common ground. Also, voltage differences between buildings are rendered harmless.

Signal attenuation in copper wiring and coaxial cable increases with frequency. By contrast, signal attenuation is flat (zero) even beyond 1 gigahertz in fiber-optic links.

Interconnection Types

There are two types of interbuilding connections. These are close grouped and remote. The closed group includes peripherals such as printers, card readers and others that are within 50 ft of the computer. These units require only simple metallic cable for interface. The environment here is usually controlled.

Remote peripherals may be located thousands of feet from the central processing unit and in or around an electrically hostile environment. Remote terminals operate in the 19.2-kbps range. Interference is usually a problem.

With remote peripherals there are three basic methods utilized for interconnection. The first is direct copper cable wiring. This is the most efficient and economic method for short distances up to approximately 50 ft. The second interconnection for remote use is the line driver. The line driver is used because of signal distortion, which limits speed and distance that is inherent in the electrical characteristics of cables. Pulse rounding of the square waves' edges are present in extended distances. All this leads to marginal reception and lost bits. Line drivers operate at 19.2 kbps; the range is between 100 ft and over 1 mile. The third interconnection method is the limited-distance modem (LDM). These are simpler versions of the conventional telephone modem. The LDM operates at 19.2 kbps to 1 Mbps at distances up to 10 or 12 miles.

Besides these three interconnections, fiber-optic cable is used as an interface to connect the peripherals to the computer. Fiber-optic cable is interfaced to standard signal RS-232C cable. These are plug-compatible to most computers and their peripherals.

Bandwidth Versus Distance

For twisted-pair wiring the bandwidth distance parameter is 1 MHz/km. For coaxial cable the bandwidth distance parameter is 20 MHz/km. For fiber-optic cable the bandwidth distance parameter is 400 MHz/km.

FIBER-OPTIC DIGITAL DATA LINK

To illustrate typical applications of the fiber-optic digital data link, the author has chosen the Canoga Data Systems CRS-100 Series Modem. A photograph of the modem is shown in figure 12–2.

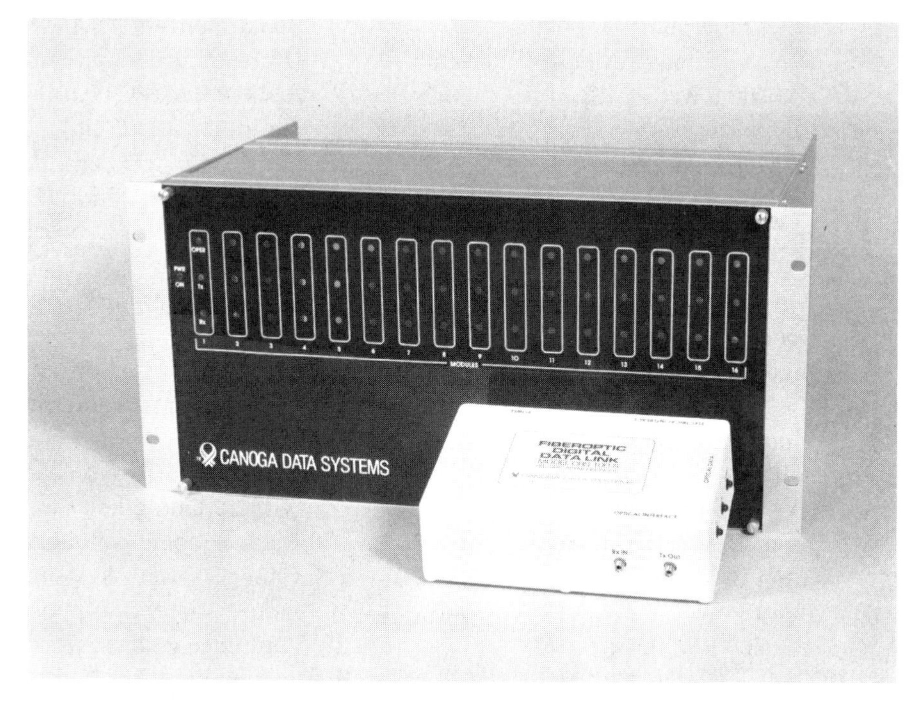

Figure 12–2 Typical Fiber-Optic Digital Data Link *(Courtesy Canoga Data Systems)*

The modem is either a stand-alone unit with a built-in power supply or is mounted in a rack as shown in the photograph. The rack is in a standard 19-in. chassis and has position for 16 of the modems. The power supply is built into the rack.

The modem is electrically interfaced as in the EIA standard RS-232C type D interface. Outputs and inputs are designated as follows:

Outputs: Voltage into 3 to 7 kΩ

+5 to +12 V (Logic 0 for data, ON or True for control signals)

−5 to −12 V (Logic 1 for data, OFF or False for control signals)

Impedance with power off: Greater than 300 Ω

Inputs: Maximum voltage: ±25 V

Impedance: 3 to 7 kΩ; less than 2500 pF (picofarads)

Specifications for the CRS-100 Series Modem

Optical Transmitter

Output Wavelength
820±15nm

Typical Output Power at Connector
200 μW (−7 dBm)

Emitter Type
Infrared LED

Optical Connector
SMA or equivalent

Optical Receiver

Input Wavelength
820±15nm

Minimum Input Sensitivity (10^{-9} BER)
−35 dBm

Input Detector
PIN diode

Input Power Overload Protected
200 μW (max.)

Optical Connector
SMA or equivalent

Physical Characteristics

Size
Stand alone—7.4×4.7×2.1 in. (18.8×11.9×5.3 cm)
Rack mount (PC board)—7.1×4.4 in. (18×11.2 cm)

Weight
Stand alone—2.75 lb (1.25 kg)
Rack mount (PC board)—1.5 lb (0.68 kg)

Operating Temperature
0° to 50°C

Humidity
0 to 50% (noncondensing)

Mounting
Stand alone—Nonskid feet with mounting holes
Rack mount—mounts into Canoga Data Systems rack mount chassis supported by card edge guides

Power Supply
Stand alone—internal

Rack mount—supplied within mounting rack

System
Transmission
Silica fiber-optic cable
Transmission Distance
3000 ft max. (1 km)
Mode of Operation
Full-duplex asynchronous (simplex operation also available)
Typical BIT Error Rate
Better than 10^{-9}
Transmission Loss
28 dB max. (coupling, splicing, and cable)

Electrical Signal Input/Output
Interface Format
EIA RS-232C (Type D)
Baud Rate
0 to 56 kbps
Minimum I/O Pulse Width
17.8 μs
Input Impedance
3000 to 7000 Ω,
less than 2500 pF
Output Impedance
Greater than 300 Ω
Maximum Input Voltage
± 25 V
Duty Cycle
0 to 100%
Input/Output Connector
Standard RS-232C subminiature D 25-pin

Digital Simplex Application (See Figure 12–3)

Probably the simplest of the fiber-optic link application is a unidirectional link between a CPU and several peripheral units. In the illustration a plotter, a printer, a card reader, and a paper tape reader are connected unidirectionally to the CPU. The plotter and printer are tied to receiver modems. The card reader and paper tape reader are linked to transmitter modems. Note that there is only one-way operation with the simplex operation.

Handshaking Operation

The handshaking protocol ensures that the terminal device connected to one data link module will be allowed to transmit, only if the terminal device connected to the other data link module is ready to receive.

The data link modules transmit status information (multiplexed with data during data transmission) on the optical cables whenever they are powered on, in contrast to modems, which must interrupt data transmission to allow one modem to interrogate the other and receive a reply. This

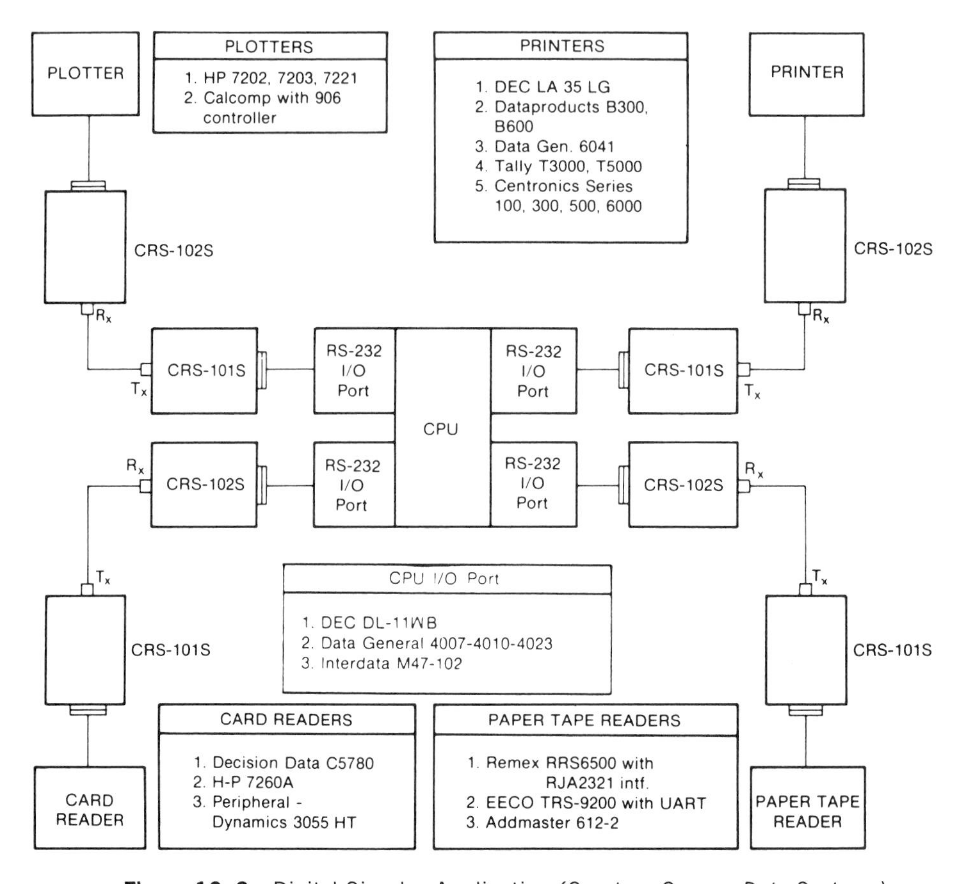

Figure 12–3 Digital Simplex Application *(Courtesy Canoga Data Systems)*

continuous monitoring protects against loss of data due to failure of the optical cables or data link module, or the receiving terminal device being not ready to receive.

Single Terminal Duplex Application (See Figure 12–4A)

When two-way communication is required, the duplex stand-alone units are utilized. In the figure, a pair of duplex units are used with transmit and receive functions linked with optical cable. The units are then tied to a terminal on one end and the CPU on the other.

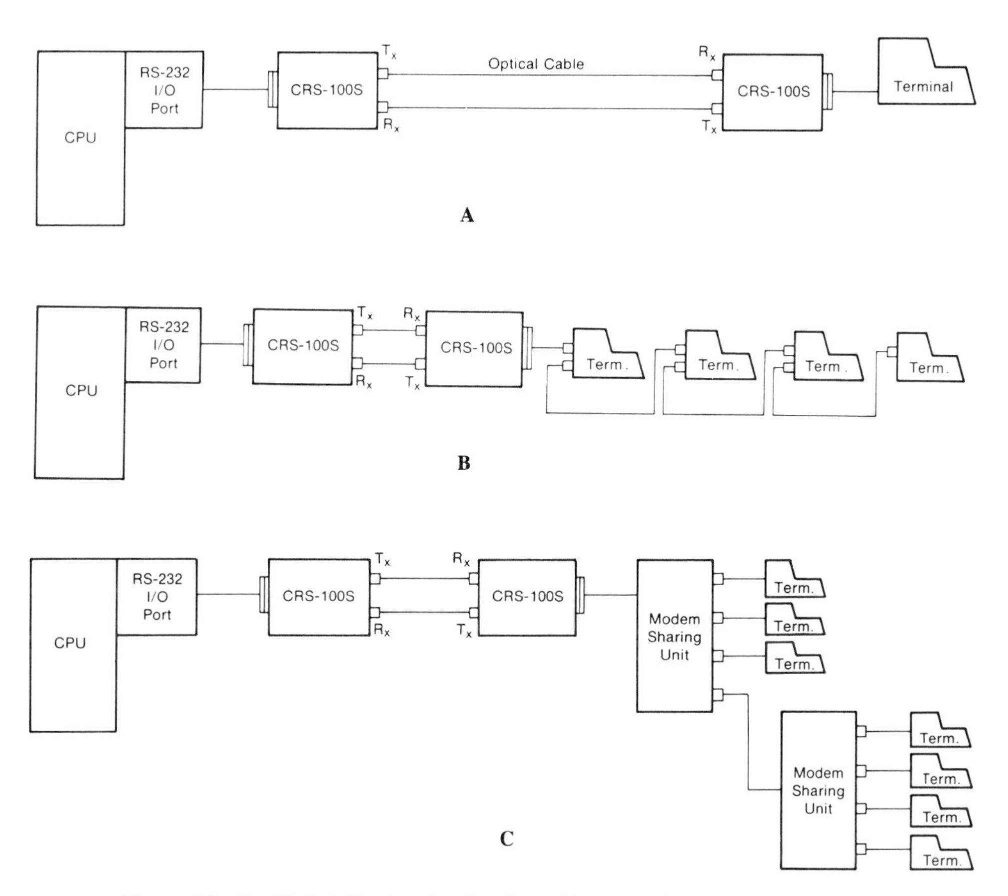

Figure 12–4 Digital Duplex Applications *(Courtesy Canoga Data Systems)*

Multiple Terminal Duplex Application (See Figure 12–4B and C)

The same communications link can serve several terminals. If the terminals are located within 50 feet of a common point and messages are short or infrequent, satisfactory operation is possible with only one terminal transmitting or receiving at one time. This can be accomplished in two different ways.

The first (figure 12–4B) is a straight polling configuration. Terminals are connected in a series-parallel daisy-chain arrangement with one terminal connected to the data link. Each terminal has a unique address assigned internally, will receive only messages preceded by its address, and will transmit only when an interrogating message preceded by its address is received.

The second method of using multiple terminals on a single communications channel is shown in figure 12–4C. These units connect the first terminal that issues a Request to Send to the data link by returning it a Clear to Send from the data link.

Other terminals requesting to send while another is transmitting are kept waiting (Clear to Send inhibited) until the terminal currently transmitting drops its Request to Send, at which time the next terminal's Transmitted Data and Request to Send outputs and Clear to Send input are connected to the data link. Received data, with an address as in the polling scheme described above, are broadcast to all terminals, but received by only one. Modem-sharing units typically have four or eight terminal ports, but can be cascaded as shown in the figure 12–4C. All data links are, of course, connected by optical cables.

CPU-to-CPU Application (See Figure 12–5)

Communication between CPUs is almost always two-way, but if one-way is desired, simplex links can be used if the system can tolerate the possibility that one CPU may send data while the other CPU is not ready to receive.

Multiplexer Application (See Figure 12–6)

When several terminals must communicate to a CPU at low to moderate data rates, a multiplexer is used. The multiplexer is interfaced to a pair of

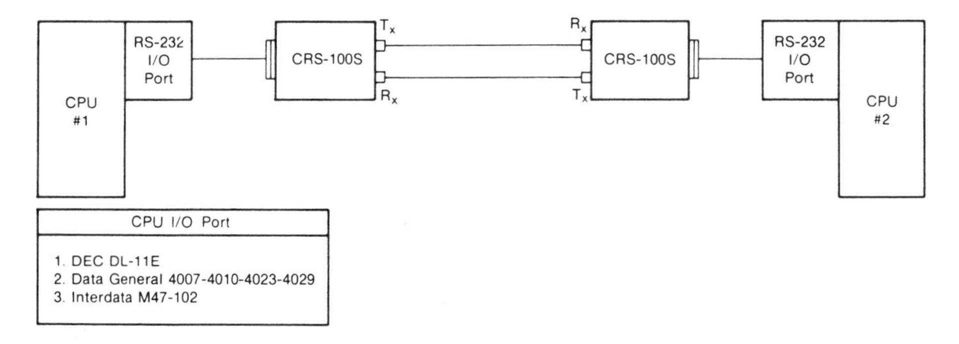

Figure 12–5 CPU-to-CPU Digital Data Link *(Courtesy Canoga Data Systems)*

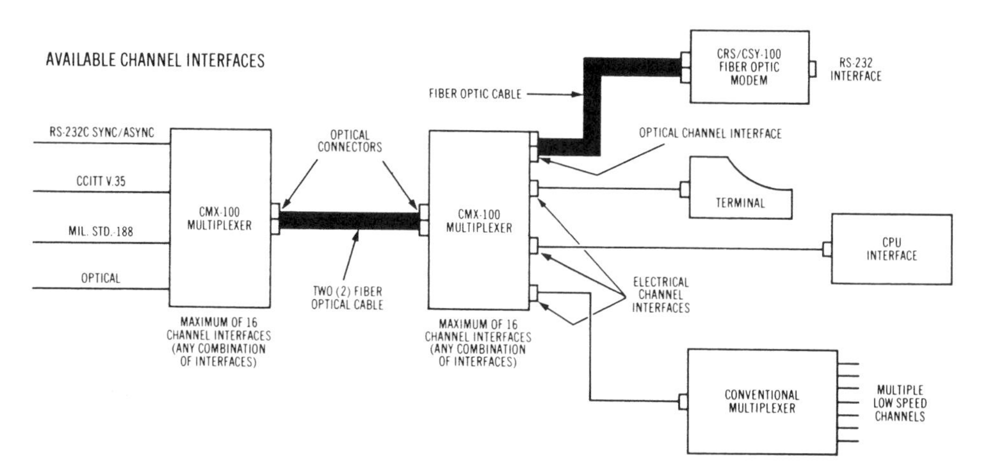

Figure 12–6 Multiplexer Applications *(Courtesy Canoga Data Systems)*

data links as shown in the figure. The multiplexer's purpose is to accept synchronous and asynchronous inputs from several sources and time divide them so they will be acceptable at the CPU and peripherals. These signals may be similar or in mixed configuration. Not only are transmit and receive data lines active, but also function information such as Request to Send, Clear to Send, Data Set Ready, Data Clocks, and so on.

Time-division multiplexers are equipped for real-time transmission. Information is not stored in memory banks. Each channel operates as if it has individual data lines, for there is no waiting or interaction with other channels.

Typical of the multiplexers on today's market is the Canoga Data System Model CMX-100. (See figure 12–7.)

Specifications for the CMX-100 Multiplexer

Optical Channel Interface

Interface Format
  Compatible with the Canoga Data Systems CRS-100 and CSY-100 Modems

Input/Output Wavelength
  820 ± 15 nm

Typical Output Power at the Connector
  200 μW (−7 dBm)

Minimum Input Sensitivity (10^{-9} BER)
  −35 dBm

Input Detector

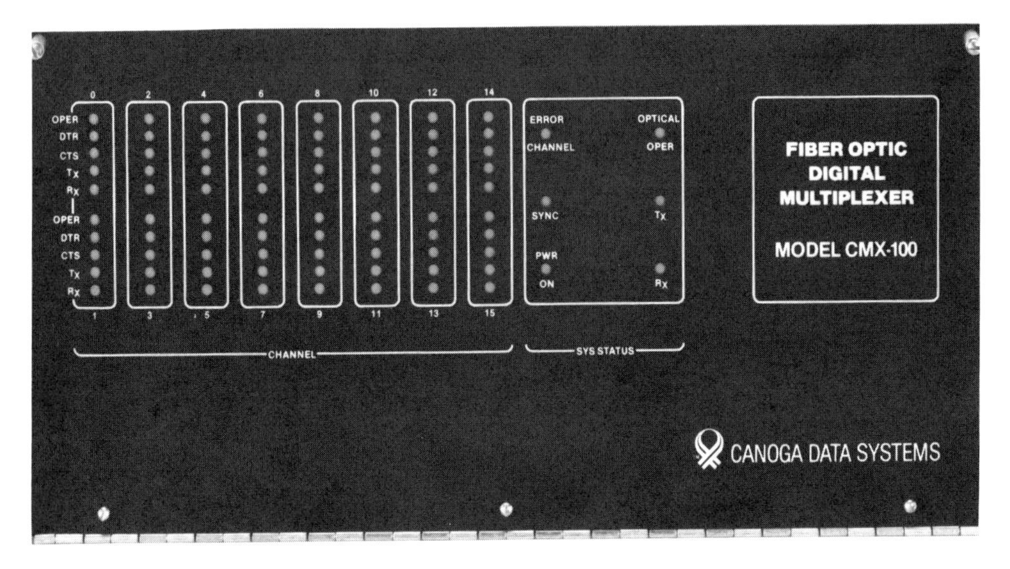

Figure 12–7 Typical Fiber-Optic Time-Division Multiplexer *(Courtesy Canoga Data Systems)*

PIN diode
Optical Connector
Amphenol Type 906 SMA or equivalent

Multiplexer Output I/O Interface
Optical Transmitter
Output Wavelength
830 ± 20 nm
Typical Output Power at the Connector
200 μW (−7 dBm)
Emitter Type
Infrared LED
Optical Connector
SMA or equivalent

Optical Receiver
Input Wavelength
830 ± 20 nm
Minimum Input Sensitivity (10^{-9} BER)
−35 dBm
Input Detector
PIN diode
Optical Connector
SMA or equivalent

Physical Characteristics
Size
Rack mount—19 × 8.75 × 14.1 in. (48.3 × 22.2 × 35.8 cm)
Desk mount—20.04 × 9.53 × 17.9 in. (51.8 × 24.2 × 45.5 cm)
Weight
Rack mount—less than 16 lb (7.3 kg)
Desk mount—less than 25 lb (11.34 kg)
Operating Temperature
0° to 50°C

Humidity
0 to 50% (noncondensing)
Power Supply
115 V, 60 Hz, less than 3 A (Redundant supply also available)

System
Transmission
Silica fiber-optic cable
Transmission distance
3,000 ft max. (1 km)
Mode of Operation
Full duplex
Typical BIT Error Rate
10^{-9} or better
Transmission Loss (Optical)
28 dB max. (coupling, splicing, and cable)
Channel Configuration
16 channels max. (in increments of two channels)
Channel Mode of Operation
Asynchronous or synchronous

Electrical Channel Interface
Interface Format
EIA RS-232C (Type D) asynchronous or synchronous (switch selectable)

Note:
Other electrical interfaces, such as MIL-STD-188, and CCITT V.35, are also available. Consult factory for specification.

Baud Rate
Async: dc to 19.2 kbps
Sync: External Clock—dc to 56 kbps
Internal Clock—1200, 2400, 4800, 9600, 19,200, and 57,600 bps (switch selectable)
Optional: instead of 57,600 bps, other data rates can be provided, such as 38,200, 40,800 bps, or 56,000 bps (Specify when ordering.)
Minimum I/O Pulse Width
17.3 μs
(Synchronous mode)
Input Impedance
3000 to 7000 Ω,
less than 2500 pF
Output Impedance
300 Ω minimum
Input Voltage
±25 V max.
Output Voltage
Compatible with EIA RS-232C spec.
RTS/CTS Delay
0, 8, 50, and 150 ms ±20% (switch selectable)
Input/Output Connector
Standard RS-232C
Subminiature D 25-pin

REFERENCES

Technical information and illustrations for chapter 12 were supplied by Canoga Data Systems, Canoga Park, California. All copyrights © are reserved.

Bibliography

Books

Allen, W. B., *Fiber Optics*. Plenum Press, London, 1979.

Clarricoats, P. J. B., Ph.D. *Optical Fiber Waveguides*. Peter Peregrinus Ltd., Institution of Electrical Engineers, Southgate House, Stevenage, Herts. SG1 1HQ, England, 1975.

Gloge, Detlaf, ed., *Optical Fiber Technology*. Institute of Electrical and Electronics Engineers, Inc., New York, 1976.

Midwinter, John E., *Optical Fibers for Transmission*. John Wiley & Sons, New York, 1979.

Pamphlets

Anderson, John D., "Fiber Optic Data Bus," Northrup Corp., Hawthorne, California.

Bark, P. R., and Wey, R. A., "Field Splicing of Optical Cables on Fiber Assessment with Optical Time Domain Reflectometry," Siecor Optical Cables Inc., Horseheads, New York.

Bielowski, W. B., "Low Loss Waveguide Communication Technology," Corning Glass Works, Corning, New York.

Bishop, O. L., and Smith, J. C., "Installation of a Fiber Optic System in an Electrical Power Station," ITT Telecommunications, Raleigh, North Carolina.

Borsuk, Les, "Fiber Optic Interconnections," ITT Cannon Electric, Santa Ana, California.

Borsuk, Les, "Introduction to Fiber Optics," ITT Cannon Electric, Santa Ana, California.

Bowen, Terry, "Low Loss Connectors for Single Optical Fibers," AMP Inc., Harrisburg, Pennsylvania.

Ebhardt, C. A., "A T1 Rate Fiber Optic Transmission System for Electrical Power Station Communications," ITT Transmission Division, Raleigh, North Carolina.

Ebhardt, C. A., Edward, A. K., and Magner, R. G., "A T3 Rate, Repeatered Fiber Optic Transmission Systems," Raleigh, North Carolina.

Eppes, T. A., Goell, J. E., and Gallenberger, R. J., "A Two Kilometer Optical Fiber Digital Transmission System for Field Use at 20 Mb/s," ITT Electro-Optical Products Division, Roanoke, Virginia.

Fenton, Kenneth, "Fiber Optic Connectors," ITT Cannon Electric, Santa Ana, California.

Martinet, R. L., "A T3 Rate, Repeatered, Fiber Optic Transmission System," ITT Electro-Optical Products Division, Roanoke, Virginia.

Miskovic, Edward J., "Fiber Optic Data Bus," ITT Cannon Electric, Santa Ana, California.

Nemhauser, Robert I., Alexander, Gil, and Duda, Richard, "Radiometry and Photometry Once Over Lightly," United Detector Technology, Inc., Santa Ana, California.

Olszewski, J. A., Foot, G. H., and Huang, Y. Y., "Development and Installation of an Optical-Fiber Cable for Communications," General Cable, Union, New Jersey.

Schumacher, William L., "Fiber Optic Connector Design to Eliminate Tolerance Effects," AMP Inc., Cherry Hill, New Jersey.

Wey, Robert A., and Cole, James A., "Optical Fibers, Cables, and Links for Computer Systems Applications," Siecor Optical Cables Inc., Horseheads, New York.

Glossary

Acceptance angle The critical angle, measured from the core centerline, above which light will not enter the fiber. It is equal to the half-angle of the acceptance cone.

Acceptance pattern The acceptance pattern of a fiber or fiber bundle is a curve of input radiation intensity plotted against the input (or launch) angle.

Active medium A collection of atoms excited to produce a population inversion.

Angstrom (Å) A wavelength value equal to 1/10,000 micrometer or 10^{-10} meter.

Antinodes The maximum amplitude point in standing waves.

Atomic lifetime The amount of time that an atom stays in an excited state before spontaneously falling to another, lower-energy state.

Beam A slender shaft or stream of light or other radiation.

Beamwidth (θ_0) The difference between the angles at which radiant intensity I (θ) is 50% (unless otherwise stated) of the peak value I_p.

Branch That portion of a cable or harness that breaks out from and forms an arm with the main cable or harness run.

Breakout The point where a branch meets and merges with the main cable or harness run or where it meets and merges with another branch.

Brewster window Glass windows placed at exact angles on the ends of a plasma tube.

Brightness Radiance. The amount of power radiated by a laser beam.

Buffer A protective material that covers and protects a fiber. The buffer has no optical function.

Bundle A number of fibers grouped together.

Bundle jacket The material that is the outer protective covering common to all internal cable elements.

Cable A jacketed bundle or jacketed fiber in a form that can be terminated.

Cable assembly A cable that is terminated and ready for installation.

Cable core The portion of a cable contained within a common covering.

Cable jacket The material that is the external protective covering common to all internal cable elements.

Cable or harness run That portion of a branched cable or harness where the cross-sectional area of the cable or harness is the largest.

Circularly polarized A wave is said to be circularly polarized when the electrical field traces out a circle in a plane perpendicular to the direction of wave propagation.

Cladding A sheathing or cover of a lower refractive index material intimately in contact with the core of a higher refractive index material. It serves to provide optical insulation and protection to the total reflection interface.

Cladding mode stripper A material applied to the fiber cladding that provides a means for allowing light energy being transmitted in the cladding to leave the cladding of the fiber.

Coherence Light waves which are in step, in phase, and monochromatic.

Continuous wave (CW) A laser whose output is continuous with no interruptions.

Core The high-refractive-index central material of an optical fiber through which light is propagated.

Corpuscle theory Theory of light in seventeenth century that suggested light consists of corpuscles that had an internal vibration of their own or in some way were controlled by waves or vibrations of the medium (such as air) through which they travel.

Coupler An optical device used to interconnect three or more optical conductors.

Coupling loss The total optical power loss within a junction, expressed in decibels, attributed to the termination of the optical conductor.

Critical angle The maximum angle at which light can be propagated within a fiber. Sine critical angle equals the ratio of the numerical aperture to the index of refraction of the fiber core.

Crosstalk Measurable leakage of optical energy from one optical conductor to another.

Crystalline Solid material that has a lattice crystal structure such as the ruby.

CVD (chemical vapor deposition) fiber A process by which a heated gas produces an oxide deposit to fabricate a glass fiber preform. The deposited glass becomes the core.

Decibel The standard unit for expressing transmission gain or loss and relative power levels. A decibel equals 10 times the log of the ratio of the power out to the power in. [$dB = 10 \log_{10} (P_0/P_1)$.]

Detector A device that converts optical energy to electrical energy, such as a PIN photodiode.

Diffraction Breaking up of light rays into bands when a ray is deflected by an opaque object or a slit.

Diffusion A scattering of light rays. A softening of light.

Directionality Not spread out in all directions. A narrow cone projected in one direction.

Dispersion The "spreadout" or "broadening" of a light pulse as it propagates through the optical conductor. Dispersion increases with length of conductor and is caused by the difference in ray path lengths within the fiber core.

Divergence Spreading out of a light beam.

Electroluminescence Electrical energy application to light-sensitive material causing emission of light (radiant) energy.

Emission Ejection of electrons from a surface by radiation.

Emission theory Theory of light developed about 500 B.C. by Greek thinkers. The theory suggested that all objects emit something which, when entering the eye, was caused by some part of the eye, and that therefore the brain could see.

End finish Surface condition at the optical conductor face.

Excitation mechanism Source of energy which excites the atoms in an active medium.

Exit angle The angle between the output radiation vector and the axis of the fiber or fiber bundle.

Extended source Light projecting from several points in space to illuminate a point from many directions.

Extinction A measure of how well a beam is polarized. It is the ratio of the power transmitted through an aligned polarizer and power transmitted through a polarizer rotated 90°.

Feedback Coupling of energy from output back into an active medium.

Fiber A single discrete optical transmission element, usually comprised of a fiber core and a fiber cladding.

Fiber buffer The material that surrounds and is immediately adjacent to a fiber that provides mechanical isolation and protection. (*Note:* Buffers are generally softer materials than jackets.)

Fiber bundle A consolidated group of single fibers used to transmit a single optical signal.

Fiber cable A cable composed of a fiber bundle or single fiber, strength members, and a cable jacket.

Fiber cladding That part of a fiber surrounding the core of the fiber and having a lower refractive index than the core.

Fiber core That part of a fiber having a higher refractive index than the cladding that surrounds it.

Fiber jacket The material that is the outer protective covering applied over the buffered or unbuffered fiber.

Fiber optics (FO) A general term used to describe the function where electrical energy is converted to optical energy, then transmitted to another location through optical transmission fibers, and converted back to electrical energy. Although the bulk of applications is in digital transmission data, fiber optics also can be applied to analog systems.

Fiber sheath A general term used variously to mean cladding, buffer, or jacket. Not to be used in military FO specifications.

Finite Measurable.

FO An abbreviation for fiber optics.

Footcandle Measure of intensity of illumination at all points of an illuminated surface 1 foot from a 1-candle source.

Frequency The number of times a wave occurs in a period of time.

Fresnel reflection losses The losses that are incurred at the optical conductor interface due to refractive-index differences.

Fringe A pattern of dark and light bands created by shining a monochromatic beam of light on a surface such as in thin films.

Fundamental mode A transverse electromagnetic mode (TEM) that oscillates in one mode only.

Gas loss A power loss, expressed in decibels, due to the deviation from optimum spacing between the ends of separable optical conductors.

Graded-index fiber A fiber whose index of refraction decreases with increasing radial distance from the center of the core.

Ground state The lowest energy state of an atom.

Harness A construction in which a number of multiple-fiber cables or jacketed bundles are placed together in an array that contains branches. A harness is usually installed within equipment or airframe and mechanically secured to that equipment or airframe.

Harness assembly A harness that is terminated and ready for installation.

Hertz (Hz) Cycles per second of a radiated frequency.

Heterochromatic Of, having, or consisting of many colors.

Incoherent Light waves that are not in step, not in phase, and nondirectional.

Index-matching materials Materials used in intimate contact between the ends of optical conductors to reduce coupling losses by reducing Fresnel loss.

Index of refraction The ratio of the speed of light in a vacuum to the speed of light in a material. When light strikes the surface of a transparent material, some light is reflected while some is refracted.

Infinite Lacking limits, extending beyond the bounds of measurement or comprehension.

Infrared Band of light wavelengths too long for response by a human eye.

Kilohertz One thousand cycles per second of a radiated frequency.

Lambda (λ) Greek letter depicting wavelength. Mathematically, $\lambda = c/f$, where c is velocity in meters (300,000,000 meters per second or 186,000 miles per second) and f equals frequency.

Lambertian A radiance distribution that is uniform in all directions of observation.

Laser Light amplification by stimulated emission of radiation.

Lateral loss A power loss, expressed in decibels, due to the deviation from optimum coaxial alignment of the ends of separable optical conductors.

Launch angle The angle between input radiation vector and the axis of the optical conductor.

Linearly polarized A wave is said to be linearly polarized when the electrical field oscillates along a line in a plane that is perpendicular to the direction of wave propagation.

Lumen Measure luminous flux of 1 square foot of spherical surface 1 foot from a 1-candle source.

Maser Microwave amplification by stimulated emission of radiation.
Megahertz One million cycles per second of a radiated frequency.
Meridial rays The rays of light that propagate by passing through the axis of the fiber and travel in one plane.
Metastable state Energy states of atoms that have extremely long atomic lifetimes.
Micrometer (mm) A wavelength value equal to 10,000 angstroms or 10^{-6} meter.
Microwave Electromagnetic waves extending from 300 MHz to 300,000 MHz.
Monochromatic Of, having, or consisting of one color.
Multiple-bundle cable assembly A multiple-bundle cable terminated and ready for installation.
Multiple-fiber cable assembly A multiple-fiber cable that is terminated and ready for installation.
Nodes The zero-amplitude points in standing waves.
Numerical aperture (NA) The characteristic of an optical conductor in terms of its acceptance of impinging light. The sine of the acceptance cone half-angle equals the square root of the difference of the square of the index of refraction of the fiber core minus the square of the refractive index of the fiber cladding.
Numerical aperture, NA (10% intensity) The numerical aperture of a fiber or bundle is defined by: NA (10% intensity) = sine θ', where θ' is the angle where the measured intensity of radiation is 10% of the maximum measured intensity when plotting either the acceptance or radiation pattern.
Numerical aperture, NA (materials) The numerical aperture of a fiber or bundle is defined by: $NA = (n_1^2 - n_2^2)^{1/2}$, where n_1 and n_2 are the fiber core and cladding refractive indexes, respectively, for step-index fibers. For graded-index fibers n_1 is the maximum index in the core and n_2 is the minimum index in the clad.
Numerical aperture, NA (90% power) The numerical aperture of a fiber or bundle is defined by: NA (90% power) = sine θ, where θ is the angle between the axis of the output cone of light and the vector coincident with the surface of a cone that contains 90% of the total output radiation power, or where θ is the angle between the axis of the input cone of light and the vector coincident with the surface of a cone which contains 90% of the total input radiation power.
Opaque Material that blocks light rays.
Optical conductors Materials that offer a low optical attenuation to transmission of light energy.
Optical spectrum Wavelengths that operate in the light and sight spectrum.
Optics A branch of physics dealing with light and vision.
Optoelectronics A branch of electronics dealing with light.
Packing fraction (PF) The ratio of the active core area of a fiber bundle to the total area at its light-emitting or receiving end.
Panchromatic Sensitive to light of all colors.
Peak radiant intensity (I_p) The maximum value of radiant intensity, I.
Peak wavelength (λ_p) The wavelength at which the radiant intensity is a maximum. Unit: nanometers (nm).
Period The length of time designated to count the number of wave cycles. Usually the time is in seconds.
Phase Value of a wave in relation to another or to a specific time.
Photoconduction Change of resistance in a material when it is exposed to radiation.
Photoelectric Electrical effects caused by light or other radiation.
Photoelectrons Electrons that absorb light energy or photons.
Photoemission Light releasing electrons from a surface.
Photon Tiny pulses of light energy. An atom stimulated in an excited state emits photons of energy. A quantum of light energy.
Photovoltaic Generation of voltage with the

use of dissimilar materials and light-sensitive material in response to radiation.

Plasma Ionized gas.

Plastic-clad silica fiber (PCS) A fiber composed of a silica-glass core with a transparent plastic cladding.

Point source Light projecting from one point in all directions.

Population inversion Movement of an atom from a ground energy state to an excited energy state.

Power density Irradiance.

Pulsed A laser whose output power has predetermined changes with time.

Quantum Unit of light energy dependent on wavelength.

Radiance The radiant flux per unit solid angle and per unit surface area normal to the direction considered. The surface may be that of a source, detector, or any other surface intersecting the flux.

Radiant energy Energy generated by the frequency of radiation of electrons. The energy of light.

Radiant intensity ($I = d\phi/d\omega$) The radiant power per unit solid angle ω in the direction considered measured in watts per steradian (W/sr).

Radiant power (ϕ) The time rate of flow of electromagnetic energy, measured in watts (W).

Radiant power ratio R is defined as the radiant power ratio ϕ_2/ϕ_1, where ϕ_1, and ϕ_2 are the measured power before and after specimen conditioning, respectively.

Radiation Process by which energy in the form of rays is sent through space from atoms and molecules as they undergo internal change.

Radiation pattern The radiation pattern is a curve of the output radiation intensity plotted against the output angle.

Ray A line of light extending between two points.

Ray angle The angle between a light ray and a reference line or plane.

Receiver optical An electro-optical module that converts an optical input signal to an electrical output signal.

Refraction The bending of a light ray when the ray moves through the intersection of two media, such as air to water.

Repeater A device that converts a received optical signal to its electrical equivalent, reconstructs the source signal format, amplifies, and reconverts to an optical output signal.

Selenium Metal whose resistance is effected by light radiation.

Skew rays Rays of light that do not propagate through the axis of the fiber.

Source The source of radiant energy, such as a light-emitting diode (LED).

Spatial coherence Phase correlation of two different points across a wavefront at a specific moment in time.

Spectral bandwidth (λ_{BW}) The difference between the wavelengths at which the radiant intensity I (λ) is 50% (unless otherwise stated) of the peak value I_p.

Spectrum Ranges of frequencies or wavelengths.

Splice Nonseparable junction joining optical conductor to optical conductor.

Spontaneous emission When an atom falls from an excited state to a ground state and in doing so emits a photon of energy, the photon is said to have been spontaneously emitted.

Sputtering Atoms jarred loose from a cathode by bombardment of ions in a plasma tube.

Standing waves A pattern created by waves reflecting back and forth within a laser cavity. These waves have the appearance of standing still if they are of the same frequency, amplitude, and moving in opposite directions.

Step-index fiber A fiber in which there is an abrupt change in refractive index between the core and cladding along a fiber diameter.

Stimulated absorption When an atom absorbs a photon and moves to a higher energy state.

Stimulated emission When an external source of power causes an atom to move from an excited state to a lower state of energy.

Tactile theory Theory of light accepted by Greek thinkers about 500 B.C. It asserted that eyes sent out invisible antennae and were thus able to feel or sense those things too distant to touch.

Temporal coherence Phase correlation of waves at a point in space at two instants of time.

Translucent Material that allows light to partially pass through.

Transmission mirror Mirror on the output of a laser that partially transmits the light beam and reflects the rest of the light back into the active medium.

Transmitter optical An electro-optical module that converts an electrical input signal to an optical output signal.

Transparent Material that clearly allows light to pass through.

Transverse wave A wave whose amplitude is perpendicular to its direction of propagation.

Ultraviolet Band of light wavelengths too short for response by a human eye.

Uniform Lambertian A Lambertian distribution that is uniform across a surface.

Unpolarized A wave is said to be unpolarized when the electrical field oscillates randomly in a plane perpendicular to the direction of propagation.

Velocity Rate of change of position in relation to time.

Visible light Electromagnetic wavelengths that can be seen by the human eye, ranging from 380 to 770 nanometers.

Wave The motion of light as it travels through space.

Wave crests The part of a light wave that has maximum vibration.

Wavefront The forward side of a light wave.

Wavelength The amount of space occupied by the progression of an electromagnetic wave.

Wave trough The part of a light wave that has minimum vibration.

Index